Challenges to American National Security in the 1990s

ISSUES IN INTERNATIONAL SECURITY
Series Editors: Robert E. Pendley and Joseph F. Pilat

BEYOND 1995: The Future of the NPT Regime
Edited by Joseph F. Pilat and Robert E. Pendley

CHALLENGES TO AMERICAN NATIONAL SECURITY IN THE 1990s
Edited by John J. Weltman, Michael Nacht, and George H. Quester

EUROPEAN SECURITY IN THE 1990s: Deterrence and Defense after the INF Treaty
Walter Laqueur and Leon Sloss

A Continuation Order Plan is available for this series. A continuation order will bring delivery of each new volume immediately upon publication. Volumes are billed only upon actual shipment. For further information please contact the publisher.

Challenges to American National Security in the 1990s

Edited by
John J. Weltman
The Johns Hopkins University
Washington, D.C.

Michael Nacht
and
George H. Quester
University of Maryland
College Park, Maryland

Foreword by
Robert R. Bowie

PLENUM PRESS • NEW YORK AND LONDON

Library of Congress Cataloging-in-Publication Data

Challenges to American national security in the 1990s / edited by John
J. Weltman, Michael Nacht, and George H. Quester.
      p.   cm. -- (Issues in international security)
   Includes bibliographical references and index.
   ISBN 0-306-43858-5
   1. United States--National security.   I. Weltman, John J.
II. Nacht, Michael. III. Quester, George H. IV. Series.
UA23.C49 1991
355'.033073--dc20   91-16332
                                                                   CIP

355.033
C4372
c.2

This series of books has been prepared as an account of work sponsored by the Los Alamos National Laboratory. Neither Los Alamos National Laboratory, any agency thereof, nor any of their employees makes any warranty, expressed or implied, or assumes any legal liability or responsibility for the accuracy, completeness, or usefulness of any information, apparatus, product, or process disclosed, or represents that its use would not infringe privately owned rights. Reference herein to any specific commercial product, process, or service by trade name, mark, manufacturer, or otherwise does not necessarily constitute or imply its endorsement, recommendation, or favoring by the United States Government or any agency thereof. The views and opinions of authors expressed herein do not necessarily state or reflect those of the United States Government or any agency thereof.

© 1991 Plenum Press, New York
A Division of Plenum Publishing Corporation
233 Spring Street, New York, N.Y. 10013

All rights reserved

No part of this book may be reproduced, stored in a retrieval system, or transmitted in any form or by any means, electronic, mechanical, photocopying, microfilming, recording, or otherwise, without written permission from the Publisher

Printed in the United States of America

The Center for National Security Studies

The Los Alamos National Laboratory's Center for National Security Studies (CNSS) investigates the complex interaction between national security policy and technological issues. The Center provides a broad perspective on policy issues related to national defense, and, specifically, promotes and conducts research and analysis on key problems in the broad areas of defense policy and arms control. A key objective of CNSS research is to assist Laboratory management and scientific staff in their decision making about technical priorities and Laboratory directions. The Center also undertakes government-sponsored research, and facilitates and coordinates Laboratory participation in public discussion and debate on matters of national security.

To carry out its programs and activities, the Center supports an interdisciplinary staff, drawing on expertise from the social sciences, including history and political science, and the physical and biological sciences. The Center has an active visitor and consultant program, which encourages short- and long-term visits by experts from government and private institutions and by university faculty members. CNSS also sponsors an active postdoctoral research and graduate research assistant program to advance the careers of promising students of national security studies.

In addition to research conducted by CNSS staff, the Center sponsors seminars, workshops, and conferences designed to stimulate broader discussion of the relationships between technology and U.S. defense policy. These activities enhance communication between Los Alamos and other organizations, such as colleges and universities, that are studying issues of interest to the Laboratory.

The publications program of the Center communicates the results of studies to Laboratory personnel and to a wider policy analysis community in government, military, and academic circles, as well as private industry. Central to the publications program is this book series, *Issues in International Security*. The volumes in the series are based on research conducted by the Center's staff and by internationally recognized experts working with the staff. A primary goal of the series is to promote the reasoned analysis of international security issues, with emphasis on how these issues shape and are shaped by technological developments.

The Authors

Steven R. David, Department of Political Science, The Johns Hopkins University, Baltimore, Maryland 21218

Lynn E. Davis, The Paul H. Nitze School of Advanced International Studies, The Johns Hopkins University, Washington, D.C. 20036

Lewis A. Dunn, Science Applications International Corporation, McLean, Virginia 22102

Harry Harding, The Brookings Institution, Washington, D.C. 20036

Michael Krepon, The Henry L. Stimson Center, Washington, D.C. 20036

Walter Laqueur, Center for Strategic and International Studies, Washington, D.C. 20006

Steven A. Maaranen, Center for National Security Studies, Los Alamos National Laboratory, Los Alamos, New Mexico 87545

Michael M. May, Lawrence Livermore National Laboratory, Livermore, California 94550

Michael Nacht, School of Public Affairs, University of Maryland, College Park, Maryland 20742

George H. Quester, Department of Government and Politics, University of Maryland, College Park, Maryland 20742

Robert W. Tucker, The Paul H. Nitze School of Advanced International Studies, The Johns Hopkins University, Washington, D.C. 20036

John J. Weltman, The Paul H. Nitze School of Advanced International Studies, The Johns Hopkins University, Washington, D.C. 20036

John Zysman, Department of Political Science, University of California, Berkeley, Berkeley, California 94720

Foreword

The decade of the 1990s offers a chance to build a new and better international order. What policy choices will this decade pose for the United States? This wide-ranging volume of essays imaginatively addresses these crucial issues.

The peaceful revolutions of 1989–1990 in the Soviet Union and Eastern Europe have swept away the foundations of the Cold War. The Eastern European nations are free; Europe is no longer divided; Germany is united. The Soviet threat to Western Europe is ending with the collapse of the Warsaw Pact and the withdrawals and asymmetrical cuts of Soviet forces. And U.S.–Soviet rivalry in the Third World is giving way to cooperation in handling conflicts, as in Iraq and elsewhere.

Much, of course, remains uncertain and unsettled. What sort of Soviet Union will emerge from the ongoing turmoil, with what political and economic system and what state structure? How far and how soon will the Eastern European states succeed in developing pluralist democracies and market economies? Are the changes irreversible? Certainly there will be turmoil, backsliding, and failures, but a return to the Cold War hardly seems likely.

The new conditions demand radical rethinking about security in the broadest sense. The task for policymaking is as challenging as it was in the immediate postwar period. Indeed it may be more so. In developing the containment strategy, the Atlantic nations were driven to cooperate by the Soviet threat in a context that was bipolar. Now the threat is absent, conditions more diffuse, and interests more diverse.

There are, however, continuities on which to build. Besides containment, U.S. and allied postwar policy was based on close cooperation among these nations and with the less-developed countries to manage their interdependence for economic well-being as well as security. Many of the practices and institutions developed for these purposes will be relevant for the future if adapted to meet the new conditions.

A new order will have to take account not only of the East-West changes but also of those among the advanced nations and in the Third World. Military threats will no doubt continue to arise, as Iraq shows. But with major East–West confrontation receding, economic relations and interdependence will move to center stage worldwide. Expanding trade, capital flows, multinational firms, and instant communications are steadily intensifying linkages and eroding national autonomy as are global problems such as the environment, climate, and drugs. In this sphere the influence of Japan and the European Community has been growing as the United States has weakened itself by a decade of profligacy.

Some of the many questions posed include:

- Can "normal" relations be achieved with the Soviet Union? Despite turmoil and "new thinking," it is still a formidable military and nuclear power. Can the military element be defused by radical arms control?
- What security regime should be sought for a stable and peaceful Europe? What role for NATO? For the Conference on Security and Cooperation in Europe? For arms control? For the European Community?
- What military strategy and forces will be appropriate for the new conditions?
- What should be the U.S. role in Europe? In the Far East? What global role should Japan and the European Community play?
- What interests do the United States and the advanced nations have in the Third World, and what commitments of resources do they require?
- How can the threats from the spread of nuclear, chemical, and other advanced weapons among smaller states be controlled?
- Can the United Nations now become a major instrument for collective security and peace-keeping, especially in the Third World?

Developing a cooperative international order will depend on achieving consensus and concerted action among many nations. That will require political leadership and many years of effort. It will also require extended analysis and debate to clarify the issues and to appraise possible ways of dealing with them. This volume is a stimulating contribution to that process.

ROBERT R. BOWIE
Dillon Professor of
International Affairs, Emeritus
Harvard University
Cambridge, Massachusetts

Preface

The Cold War, we are now told, is over. The Soviet-American relationship, which had dominated international politics for more than four decades, has been fundamentally transformed and radically reduced in importance. Has the world indeed changed this radically? Are the changes that have occurred irreversible, or could they be set aside, say, by domestic reaction in the Soviet Union? Are there no continuities that survive the dramatic events in the Soviet Union and Eastern Europe at the onset of the decade? If indeed fundamental transformations have occurred, what will take the place of the old relationships? From an American perspective, what is the context within which foreign and national security policy must be set in coming years?

The essays in this volume cast some light on the questions. They are the outgrowth of a partnership between the Center for National Security Studies (CNSS) at the Los Alamos National Laboratory and the School of Public Affairs at the University of Maryland at College Park, with the collaboration of the School's Center for International Security Studies and the University's Department of Government and Politics. Through our joint efforts a number of expert and distinguished analysts were asked to address the world that will face American policymakers in the coming decade, and to suggest the main lines that our national security policy should take. National security was deliberately construed in our endeavors in a broad sense, to cover the range of physical threats that might arise for the nation and the force structures, doctrines, and military policies mobilized to meet those threats. National security was also taken to include those internal and external factors—such as economic or technological change—that might arise and whose direct or indirect effect would be to diminish or to enhance the nation's capacity to meet physical threats.

John J. Weltman sets the stage by drawing the contrast between the continuity that might have reasonably been expected at the outset of the Bush administration, the complexity now facing it, and the likely implications of that complexity for American policy and military force structure.

Central to any assessment of the evolution of international politics is an evaluation of the Soviet–American relationship. Robert W. Tucker and Walter Laqueur examine the roots of today's Soviet external behavior and the question of the degree of its reversibility. Tucker points to the centrality of Europe and suggests reasons for caution about the prospects for fashioning a stable political settlement in that region. He stresses the importance of continued American engagement. Laqueur suggests the domestic political and ideological basis for Soviet behavior.

Steven A. Maaranen describes how arms control has influenced U.S. nuclear weapons, and how arms control along with other factors will continue to influence U.S. military strategy and the roles for nuclear weapons. Michael M. May and Michael Nacht assess the strategic military relationship between the United States and the Soviet Union. May examines the prospects for further cuts in force levels without changing current doctrinal premises. Nacht points to the predominance in the past of domestic political factors in directing the arms-control process and suggests that this will likely continue to be the case.

Lynn E. Davis examines prospects for further evolution in European political and military structures, assessing the changes in military structure and doctrine that new political developments in Europe will call forth and the role of the arms-control process in bringing the changes forward.

Harry Harding suggests an Asian region increasingly characterized by political and economic complexity, and increasingly less dominated by the Soviet–American relationship. He suggests, however, that the tensions and uncertainties which this complexity will work make some continued American presence still valuable.

Steven R. David and Lewis A. Dunn see serious continuing potential for conflict in the Third World, conflict that will continue to engage American interests. David suggests that the decline of Soviet–American hostility in Europe will not eliminate threats to our interests in the Third World and the need to fashion military instruments responsive to that threat. Dunn likewise points to the potential for the spread of advanced military technologies in exacerbating a host of regional and local conflicts and the necessity for the United States to continue to seek ways to exercise a stabilizing influence.

Michael Krepon and John Zysman analyze the economic roots of change in the international order, and the implications of that change for technology. Krepon points to the implications of a growing multipolarity in capabilities to exploit space for both military and commercial purposes. Zysman points to the economic and technological roots of the rise of new powers and suggests important implications that will arise in the military sphere from this growing economic multipolarity.

Preface

None of these changes in the external political environment will automatically produce change in American domestic policy. In a concluding essay, George H. Quester explores the domestic basis for American external commitments, factors working to induce us to reduce those commitments and those causing us to retain them.

It is a pleasant duty for the editors to acknowledge the contribution of many people essential to the production of this volume. A project of this type always involves inputs from a great number of organizations and people. The risk is that too few acknowledgments will be voiced, rather than too many.

At Los Alamos, Arthur S. Nichols, the editor of CNSS publications, established with the publisher editorial procedures and schedules, and he copyedited the volume. Also at Los Alamos, Janis Dye prepared the manuscript for the publisher with skill and unflappable good humor in the face of extremely short deadlines.

At the University of Maryland the workshops have been a portion of the activities of the Center for International Security Studies at Maryland, a campuswide research venture in issues of arms-control and defense policy. The Center, directed by Catherine Kelleher, is housed administratively in the School of Public Affairs, but includes faculty and students from all across the University. Among the students particularly involved with the workshops leading to this book have been Patricia Davis, James Derleth, Virginia Foran, Karen Ipe, Nancy Latham, and Katherine Walter. Their assistance is greatly appreciated.

Contents

1. **The Setting for American National Security in the 1990s** 1

   John J. Weltman

   Breaking with the Past 1
   The Soviet Union Retrenches 4
   European Implications 6
   East Asian Implications 9
   Third World Implications 11
   A World of Greater Complexity 12
   A Changing International System 17
   Changes in American Foreign Policy 19

2. **Some Considerations on the Soviet–American Relationship in the 1990s** 25

   Robert W. Tucker

3. **The New Thinking and Its Limits: Soviet Foreign Policy under Gorbachev** 41

   Walter Laqueur

   The New Thinking 43
   China and the Soviet Union 50
   The Third World and the Soviet Union 52
   The United States and the Soviet Union 53
   Conclusions 54

4. Arms Control and the Future of Nuclear Weapons 57
Steven A. Maaranen

The Arms-Control Context 58
Strategic Nuclear Arms Negotiations 60
Strategic Defense and Space Arms Control 64
Theater Nuclear Arms Negotiations 66
Future Nuclear Force Reductions 68
    Strategic Nuclear Arms Control 70
    Strategic Defense and Space Arms Control 72
    Theater Nuclear Arms Control 73
Conclusions 73

5. Strategic Nuclear Weapons after START 77
Michael M. May

Introduction 77
Strategic Consequences 79
    Force Survivability 80
    Target Coverage 84
    Civilian Fatalities 88
Discussion 89

6. Strategic Arms Control and American Security: Not What the Strategists Had in Mind 93
Michael Nacht

What Is Arms Control All About and Who Says So? 94
Personalities, Domestic Politics, and the Sense of History 102

7. Beyond German Unification: The West's Strategic and Arms-Control Policies 107
Lynn E. Davis

Introduction 107
Conventional Forces in Europe 109
    What Goal for Conventional Defense? 110

Defensive Strategies and the Future of Forward Defense	111
The Future Role of American Troops in Europe	113
Future Directions in Arms Control	114
Nuclear Weapons in Europe	116
Choices for Strategies and Force Postures	116
Future Directions in Nuclear Arms Control	120
Conclusions	121
The Future of NATO	121
The Strategic Purposes of Conventional Forces	122
The Role of American Troops	125
Conventional Arms Control	126
U.S. Nuclear Weapons in Europe	126
Reconstructing the Foundations of Peace	128

8. American Security Policy in the Pacific Rim 131

Harry Harding

The Elements of the Strategic Situation in East Asia	132
The Evolution of the Three Regional Disputes	132
The Rise and Retreat of the Soviet Union	134
The Economic Dynamism of the Region	138
Trends toward Multipolarity	141
Issues for American Policy	143
Forward Deployments	144
Alliance Management	145
Military Strategy	146
Arms Control and Regional Disputes	147
Diplomatic Strategy	150
Conclusions	151

9. Why the Third World Matters 153

Steven R. David

Third World Threats to American Interests	154
The Strategic–Military Threat Posed by the Third World	154
The Threat to American Economic Interests Posed by the Third World	156
The Threat to American Political–Ideological Interests Posed by the Third World	156

The Hyper-Realist Approach to the Third World 157
Responding to the Hyper-Realists 158
   The Strategic Military Threat Posed by the Third World ... 161
   The Threat to American Economic Interests Posed
      by the Third World 168
   The Threat to American Political–Ideological Interests
      from the Third World 173
A Truly Realistic Approach to the Third World 174

10. New Weapons and Old Enmities: Proliferation, Regional Conflict, and Implications for U.S. Strategy in the 1990s 179

Lewis A. Dunn

The Proliferation of Advanced Weaponry 180
   From Advanced Conventional Weapons to Weapons
      of Mass Destruction 180
   Regional Proliferation Trends 182
Advanced Weaponry, Regional Conflict, and Global
   Spillovers 187
   Patterns of Regional Conflict 187
   Global Spillovers of Proliferation 191
A Proliferation Containment Strategy 195
   Checking Further Proliferation 195
   Containing Regional Consequences and Global
      Spillovers 200
Containing the Proliferation Threat 203

11. Military and Civilians Uses of Space: Lingering and New Debates 205

Michael Krepon

Introduction 205
Lingering Debates from the 1980s 205
New Debates for the 1990s 210
Open Skies: The Policy Issues and Debates 211
The Role of the Media 212
Multilateral Verification of Peacekeeping Operations 214

12. Security and Technology 219

John Zysman

America's Changing Position in the Global Economy 220
    American Manufacturing's Declining Position 221
    A Multipolar Global Economy 225
From Spin-Off to Spin-On Technology 228
Will American Industrial Decline Reshape
   the Security Structure? 231
    The Economic Projection of Influence 232
    Security and Military Equipment 232
    Will New Players Alter the Security Configuration? 234

13. Predicting the Future of American Commitments 237

George H. Quester

Why Americans Care 239
Ethnic Considerations 241
Changes in Precedent 243
Economic Changes 244
The Proliferation of Weapons 246
Nuclear Proliferation 247
Chemical and Biological Warfare Proliferation 248
Delivery System Proliferation 250
Naval Deployments 251

About the Authors 253

Index 257

1

The Setting for American National Security in the 1990s

John J. Weltman

BREAKING WITH THE PAST

The history of international relations since World War II has been filled with proclamations of novel and radical change, yet its course has remained remarkably steady. The Soviet–American relationship has dominated the field. Foreign and defense policy for each superpower has been driven preeminently by the need to manage its relationship with the other. The relationship has never quite flared into open conflict, yet fundamental accommodation of the conflict or resolution of the issues that brought it into being has eluded the protagonists. Two recent American administrations tried and failed to break with the past. A third administration, skeptical of predictions that international relations would undergo a radical break with the past, found itself forced to cope with events that seemed to cast into doubt the primacy of the Soviet–American relationship in world affairs, and all that that primacy had implied for foreign and defense policy.

The Carter administration entered office convinced that the Soviet–American relationship had been rendered obsolete as a benchmark for American policy. Nuclear-weapons balances were irrelevant to American security because these weapons had lost all political utility. An increasing international economic interdependence dictated that military force, in general, was of declining utility

---

John J. Weltman • The Paul H. Nitze School of Advanced International Studies, The Johns Hopkins University, Washington, D.C. 20036.

in the relations of nations. Important issues would be resolved to mutual benefit or not at all. Satisfactory resolution of such issues required policies based on cooperative behavior directed to amassing a voluntary consensus among all parties. Attempts to gain advantage by conflictive behavior involving the use or threat of force would be counterproductive at best and might leave all parties worse off than before. The administration thus proclaimed novelty in international relations. The Soviet–American relationship, the *East–West* relationship, had been supplanted in importance by a *North–South* struggle over the production and distribution of wealth. In neither of these relationships was military power, as traditionally understood, especially useful.

By the close of its term the Carter administration had been forced to admit that the obsolete forces of the past still survived and exercised a powerful influence over American policy. The Soviet–American relationship reasserted itself with a vengeance. Far from resting content with the view that nuclear balances could be ignored, the administration found itself calling for a major program of strategic nuclear rearmament and for new nuclear deployment programs in Europe to offset adverse nuclear-weapons balances that had been taken as calling into doubt the credibility of the American commitment to come to the aid of its allies. The administration felt it could not ignore Soviet use of force in Afghanistan to retain that country within its sphere of dominance and indeed pointed to Afghanistan to justify the collapse of arms-control negotiations and the beginning of military rearmament.

Nor did the North–South relationship exhibit that reliance on cooperative and consensual forms of behavior that the Carter administration had anticipated would characterize it. The Iranian revolution forced the administration to grapple with the consequences of military power inadequate to protect perceived American interests. The security of the West's supplies of oil from the Persian Gulf was perceived to be endangered, and the administration reacted by proclaiming its readiness to use military force to preserve that supply, and then by moving to create the military forces that would enable it to fulfill its proclamation: forces originally designated as the Rapid Deployment Force, which has since become Central Command.

The Reagan administration took power similarly determined to break with the past. Its most dramatic departure lay in the Strategic Defense Initiative (SDI), justified as intended to remove the physical vulnerability of American territory to nuclear attack. In essence the SDI was a fundamental challenge to the nuclear deterrence posture that had lain at the heart of American military policy for thirty years or more. The administration was skeptical about the utility of arms control or other negotiations with the Soviet Union, suggesting that the ideological gulf that separated the two superpowers made the resolution of differences through negotiation a doubtful prospect. Rather than through negotiation, the administration sought to change the terms of the Soviet–American relationship through change in the American military posture.

Far from accepting any notion that North–South conflict had supplanted the East–West struggle, the Reagan administration proclaimed the centrality of the Soviet–American relationship in the Third World. Conflict in the Third World was fundamentally influenced by Soviet or Soviet-client expansionist motives. It required an American response, which was to be primarily military in character.

By the end of its term the Reagan administration, as had its predecessor, was to reach an accommodation with outside forces whose efficacy it had once denied, an accommodation that belied its premises on coming into office. The most dramatic was in nuclear arms control, where the completion of the Intermediate-Range Nuclear Forces Treaty was to eliminate a category of weapons previously proclaimed as essential to the preservation of the European alliance. Completion of the treaty was accompanied by considerable preliminary work on strategic nuclear forces arms control, which presaged an agreement that might later reduce those forces by half with the prospect of further reductions in later years. These concrete products of a negotiating process whose legitimacy the administration had earlier questioned were accompanied by the spectacle of President Ronald Reagan, at the Reykjavik summit with Soviet leader Mikhail Gorbachev, seeming to enthuse over the prospect of the abolition of nuclear weapons altogether. Defenders of the administration's consistency pointed to its original doubts about the prudence, or indeed morality, of continued reliance on a policy of nuclear deterrence employing nuclear offensive forces, and to its intention to create a new form of security in a defense-dominant world. By the administration's closing months, however, the strategic defense system that had been the centerpiece of its strategic vision at its outset had been much reduced in scope and fundamentally changed in purpose. It now seemed designed to reinforce nuclear deterrence as traditionally understood, rather than to replace it.

While the Reagan administration at its close was not to withdraw formally from its earlier assessment that Soviet-sponsored instigation lay at the heart of Third World conflict and required a military response, its policy in Central America, the archetype for its approach to the Third World, had undergone a forced moderation. The military solution that the administration had at first sought through its support for the Nicaraguan opposition had evaded it. Elsewhere the administration had come to offer support to Marxist regimes—as in southern Africa—when local circumstances suggested it might be prudent.

This recent history of retreat from proclamations of novelty in the structure of international relations, or in the capacity of change in American policy to transform the environment, should have a chastening effect on renewed assertions that the world we have known since 1945 is at a watershed. Yet the coming to power of a new administration in Washington was a convenient moment for reflections and reassessment. The papers in this volume grew out of such an ambition. Our intent was to explore the national security environment and the range of policy that the Bush administration might marshal to meet it. National security was deliberately construed in a broad sense not only to cover the range of physical threats that might arise for the nation and the force structures, doctrines, and military policies which might be mobilized to meet those threats, but also to include those internal and external factors that might arise whose direct or indirect effect would be to diminish or to enhance the nation's capacity to meet physical threat.

THE SOVIET UNION RETRENCHES

It was difficult to avoid speculation at the outset of the Bush administration that the international environment at the turn of the century would differ qualitatively from the world that has seemed so relentlessly enduring to us since 1945. Change in the Soviet Union's international role presented the most dramatic potential, although perhaps, in the end, of less significance than other prospects. There was widespread agreement that the Soviet economy faced major structural problems. There was less agreement as to the implications of these problems for Soviet external behavior. The logic of the Soviet situation suggested external retrenchment for at least the immediate future in order to sustain economic reform at home. One could indeed observe a number of trends that seemed to bear out this logic. Troop withdrawals started in Central Europe, which began to fulfill the unilateral reductions promised in Gorbachev's address to the United Nations in December 1988. A withdrawal from some foreign commitments seemed in train. Soviet combat forces left Afghanistan. The Soviet Union encouraged her Vietnamese ally to withdraw from Cambodia and her Cuban ally from southern Africa, and to seek negotiated settlements of these conflicts.

In other respects external retrenchment still seemed more promise than substance in the early days of the Bush administration. Little evidence had appeared to suggest that Soviet forces had been reoriented to an operational defensive that would render them no longer worrisome to their neighbors. Supposed changes in military doctrine to emphasize the defense had yet to be unambiguously reflected in force structure or maneuver patterns. The promised major cuts in defense expenditure had yet to be reflected in demonstrable ways in deployed formations and weapons systems, whether at the conventional or at the strategic levels. While arms-control negotiations continued at a variety of levels, it was unclear whether the Soviets were prepared to give up hard-won military advantages in order to seek balances genuinely less threatening to each side, or whether Soviet negotiating postures were instead designed primarily to exploit fissiparous tendencies in Western domestic and allied opinion.

Beyond these questions lay some broader ones. Whatever the logic flowing from her domestic economic difficulties, could the Soviet geopolitical position survive the implications of the retrenchments the Soviet Union had promised? If the Soviets did indeed carry through on these promises, should the Western alliance and the United States look to the consequences with hope or with anxiety?

The Soviet position at the end of World War II was essentially a function of the territorial gains made by the Red Army. Governments subordinate to the Soviet Union was prepared to scale down its conception of the minimum requirements for its security in Central Europe from that established forty-five years ago, to one more compatible with a reduced Soviet presence. Such a conception tee for the survival of those governments. Major Soviet military retrenchment could not be carried very far before it became incompatible with continuance of this political status quo in Central Europe. The question remained whether the Soviet Union was prepared to scale down its conception of the minimum requirements for its security in Central Europe from that established forty-five years ago, to one more compatible with a reduced Soviet presence. Such a conception would presumably trade the domestic autonomy of the Central European states for continued commitment to an international posture of support for Soviet foreign policy—or at least a commitment not to oppose that policy. Could a Soviet regime prepared to draw the lessons of retrenchment in terms of such a reduced conception of security survive domestically? Would the increasingly complex political currents of Central Europe allow such a process to take place gradually and incrementally, or would violent upheavals occur as popular expectations were raised? If the latter, could the process of retrenchment itself survive them?

If a reordering of security arrangements in Central Europe could nevertheless be brought about, should such a prospect be welcomed by the West? Although the advance of the Red Army into Central Europe in 1945, and the political arrangements imposed by that army's presence, inspired fears for the security of the remainder of Europe and thus was the motive force for the creation of the policy of containment, for the formation of the Western alliance, and for Western military rearmament, it also solved—or at least anesthetized—European problems that had previously been the cause of numerous wars and crises.

Most obviously, the Red Army's advance enforced the division of Germany. The creation of a united Germany in 1870 had led eventually to fears on the part of its neighbors that such a powerful European entity could not be contained if it sought further expansion. After 1945 the advance of the Red Army, and consequently the presence of two hostile alliance blocs meeting on the inter-German border, functioned as a guarantee for all sides that such a powerful state could not arise again. At the beginning of the new American administration, one could not predict precisely what would be the effect on the German question of a looser structure of Soviet control in Central Europe, or when. But it would certainly raise the specter of closer relations between the German states, no longer governed by the constraints of the two alliance systems. Such a major change in the political map of Europe could not but raise the level of tension and uncertainty in the region.

The Red Army's advance in 1945 put to rest another problem that had bedeviled the politics of Europe for decades. The slow decline of Austria–Hungary had raised the question of the nature of the succession to that empire in dominating the Danubian basin and southeastern Europe. This succession question largely explains why local disputes in the region tended to become invested with great-power interest and commitment. World War I destroyed Austria–Hungary, but left only a number of smaller powers in its place. In 1945, however, the outcome of the war clearly left the Soviet Union in place as Austria–Hungary's successor to dominance in the region. At some point a looser structure of Soviet control in the region would call this succession into question, reawakening the latent conflicts among and within the various states in the region and raising the possibility of renewed competitive engagement by outside powers in those conflicts.

To be sure, many involved in the national policy debate argued that Soviet retrenchment might be intended only as a breathing space to allow the Soviet economy time to develop the infrastructure to support a military establishment at qualitative levels comparable to those in the West. Under this view, once this tactical goal had been achieved the regime would be free to return to expansionist postures. Against this view, it was argued that the character of the structural changes required in Soviet society and economy necessary to support such an infrastructure are so pervasive, and the degree of dependence on outsiders which would result so extensive, that the domestic base for reversion to external adventures would no longer exist.

EUROPEAN IMPLICATIONS

Whether Soviet retrenchment was to be regarded as a short-term, reversible tactic or an irreversible and fundamental shift in external orientation, the consequences of such retrenchment are not necessarily favorable, either to Western interests or to the avoidance of potentially dangerous crises. While beginning the process of dismantling situations that gave rise to the Cold War, it may at the same time open up sources of complexity, miscalculation, and conflict long buried beneath the frozen landscape of bipolar hostility.

The nature of the Soviet regime and the character of its external behavior have driven American foreign and national security policy in the postwar period. Change in this Soviet behavior would imply important adjustments for American policy. But other developments may be postulated whose effect might well be to reduce the salience of the Soviet–American relationship, whatever its character. We need not postulate the end of the Cold War to accept that the decline of bipolarity will project us into a more complex, and inherently unpredictable, new environment.

States that formerly accepted the parameters of a security relationship fundamentally shaped by the United States are growing increasingly assertive and are demanding greater autonomy in setting their own priorities. The Third World, formerly in broad respects the object of superpower rivalry, is growing increasingly capable of resisting external intrusions and following the dictates of its own indigenous priorities. These developments may all themselves be products of fundamental change in the economic relationships that have applied in the postwar world.

Fears that the Western alliance could not survive have been with it since its origins. That it has survived these vicissitudes can lead to the complacent conclusion that it will continue to do so in fundamentally unchanged form. However, two major trends may put such a conclusion in doubt: first, the perceived decline in the threat that provided the reason for the alliance to come into existence in the first place, and the glue that has enabled it to resist numerous centrifugal temptations since then, and second, the growing economic vitality of Europe and many of its component states, which transforms the relationship of dominance and dependence that underlay the original conception of the alliance.

The Western alliance came into being to counter the threat posed by large Soviet military forces that remained in occupation of Central Europe four years after the conclusion of World War II. Fears existed that these forces might simply mount an invasion of Western Europe, although this was not the sole—or even primary—reason for concern over Soviet military forces. It was feared that the existence of these forces might coerce Western governments into adopting policies or domestic arrangements conducive to Soviet desires. The 1948 coup in Czechoslovakia seemed an example of the latter sort of danger, as had earlier situations, as in Hungary, where Communist parties were brought to power under Soviet pressure and contrary to any popular mandate. The original conception of the alliance was as an American guarantee to come to the aid of the Western Europeans should they come under military attack from the East. The guarantee itself—a break from America's isolationist past—would suffice to reassure the allies that an attack would not take place and would thus remove the potential for Soviet-inspired coercion. Only in 1950, in the wake of the outbreak of the Korean war, did it appear necessary to supplement the American political guarantee with the provision of American troops deployed permanently in Europe, and with the formation of an integrated alliance military command headed by an American officer.

What was notable in the early days of the Bush administration in Washington was the extent to which those closest to the putative military threat doubted its continued existence. The view that the Soviet Union no longer represented a military threat was most widely held in Germany. The view that the threat remained potent was widely held across the spectrum of political opinion in the United States. In previous intra-alliance conflicts it was the Europeans who demanded reassurance of the stability of American commitment—whether they feared American withdrawal, the diversion of American attention elsewhere through out-of-area adventures, or some form of superpower condominium that would leave them exposed on the continent. Here, it was the commitment of the European pillar of the alliance—West Germany—which was questioned by the United States as well as other Europeans.

Public opinion is historically volatile, and one should beware of projecting current trends into the future. One could note, however, that the relative relaxation of German opinion was not merely the result of optimism about the nature of the Soviet Union since Gorbachev's accession to power in 1985. In an important sense, these attitudes became apparent as early as the debates over intermediate nuclear force deployments at the end of the 1970s. A situation in which the alliance partner supposedly most vulnerable to danger is least concerned to meet that danger must imply that alliance relationships in Europe could not survive unchanged.

Among the potential ways in which the alliance could undergo change, by far the most likely would be a shift in responsibilities from America to Europe. In such a situation the predominant role of the United States as military leader would decline. There might be some substantial diminution of the American military presence in Europe, and alliance military policy would deemphasize its current reliance upon the threat to use American nuclear weapons in the event of a failure of conventional defense efforts. The American commitment to the defense of Europe would survive, but it would move closer to the original guarantee notion of the alliance. Some mechanism providing a greater degree of defense cooperation among the European allies themselves—or some of them—would take over many of the responsibilities now borne by the United States. The germ of such a mechanism might perhaps be seen in the revival of the Western European Union as a vehicle for defense cooperation among the Europeans, and in the nascent bilateral cooperative activities between the British and the French and between the French and the West Germans. An alliance in which the European pillar took on added responsibilities relative to the American one need not be confined to conventional forces. A European nuclear strategy could be based upon the existing national nuclear forces of Britain and France (both undergoing considerable expansion and modernization), or upon a European force that might arise from the amalgamation of these national forces.

The stability of such an arrangement would largely be a function of the allies' continued perception of threat from the East, and of the degree of uniformity of such perceptions among the European allies. A moderately threatening Soviet Union might well hold such an alliance together. A Soviet Union no longer perceived to represent a believable military danger, or about which there were widely varying estimates of danger among the allies, might well lead to fragmentation of the alliance.

Fragmentation of the alliance could conceivably take many forms, the most likely flowing from a fundamental shift in German foreign policy orientation. In an important sense, West Germany was the linchpin of the Western alliance in Europe. When the original guarantee conception of the alliance was supplanted by one that saw a need for the alliance to mount substantial military force to deter and perhaps to resist attack from the East, it was realized that such a force could be mounted only with German participation. Germany at the time remained disarmed and occupied. Furthermore, other European members of the alliance variously regarded Germany with concern as a potential threat to their own security. The bargain that was to be adopted saw West Germany rearmed and given sovereignty in return for a commitment to integrate its regained military forces into the structure of the alliance.

As has already been noted, German opinion in recent years had been noticeably more optimistic than that of most other members of the alliance about the nature of the risk from the East. The continued division of Germany was no doubt an important factor contributing to these attitudes, as was Germany's special sense of vulnerability in any conflict that erupts in Central Europe. The Soviet Union was uniquely placed toward Germany in that it was the only power that could conceivably offer the Germans reunification. This possibility was inherent in the structure of European politics, however unlikely it then appeared that the Soviet Union would actually contemplate such an offer. Thus, from the German perspective, amelioration of relations with the East inevitably carried the added attraction of improving the prospects for national reunification.

By the fall of 1990, German unification, in the form of East Germany joining the West German federal state, had occurred. The Soviet government had not only accepted this result but had also agreed that the united German state could remain a member of the North Atlantic Treaty Organization (NATO) albeit with some restrictions on the entry of non-German NATO forces into the former East German territory. The Soviet Union, furthermore, agreed to withdraw its troops from German territory by 1994. While these developments represented a triumph and vindication of the Western alliance, some questions arose with renewed emphasis: Could the NATO alliance survive success? Would the diminution of a sense of threat from the East remove the incentive for the Western allies to coordinate their foreign and defense policies? If NATO survived juridically would it continue to dominate the actual behavior of its members?

EAST ASIAN IMPLICATIONS

The potential for change in the security structure that the United States erected in the years following the conclusion of World War II was by no means confined to Europe. It may also be seen in East Asia, the other main focus of American security concerns in the postwar years. In East Asia, as in Europe, the predominant object of American concern was the Soviet Union, although in East Asia this was supplemented by a concern with Chinese expansionism, supposedly in league with the Soviet Union. This belief in Soviet–Chinese solidarity driven by a common ideology was cemented by the Chinese intervention in the Korean War. An important motive for American involvement in the Indochinese conflict was the continued perception of an expansionist China allied with the Soviet Union. Many years after this notion of a Soviet–Chinese collaboration had become, in fact, a dead letter and had been succeeded by a period of high tension between the two Communist powers, American policy adjusted to this fact in the Sino-American détente of the 1970s. Indeed, this détente assisted in the process of American extrication from Indochina. The détente with China was quickly succeeded by a period in which—from the American perspective at least—China was considered almost a part of our anti-Soviet alliance system. In recent years China had taken a position of more equal distance between the superpowers, giving rise to concerns on the American side that a renewed close relationship with the Soviet Union might again be possible.

The centerpiece of the American security structure in East Asia, however, has been Japan. As was Germany, Japan was disarmed and occupied at the end of World War II. As was Germany, Japan was later made the object of a bargain that was to form the centerpiece of the American security structure in her region. As was Germany, Japan was the object of security concerns by various states in the region—most notably those countries that had fought against or had been occupied by Japan. Many of these were former colonial possessions of European powers that had now gained independence. The bargain struck with respect to Japan differed in some important ways from that struck with respect to Germany. While the United States believed that the security threat from the Asian Communist powers required the maintenance of countervailing military power in the region, the moat represented by the Sea of Japan allowed this military force to be primarily naval and air. The creation of a large-coalition military-force structure, to which Japan would contribute, was not considered necessary. Thus, the bargain struck was one in which Japan allowed the United States to base military forces on Japanese territory indefinitely, in return for the acquisition of full sovereignty over its own affairs. Japan regained military forces, but these were to be kept limited to defense of Japanese territory. Further rearmament would have excited anxiety on the part of America's smaller allies in the region, and Japan's potential economic contribution was felt to be its principal responsibility for regional security.

As the Bush administration took office, this bargain was felt to be under threat. Unlike the situation in Europe, however, where one might argue that it was the Germans who questioned the continuance of the bargain, in East Asia it was the United States that cast doubt upon it. Relative American economic decline had led to persistent American calls on the Japanese to bear a greater part of the costs of the security structure agreed to almost forty years before. In particular, the United States has asked that the Japanese expand their military forces to meet requirements beyond that of defense of the Japanese home islands. In addition, there have been recurrent suggestions emanating from Washington that economic stringency, or displeasure with Japanese import–export policy, might cause the United States to diminish its military presence in East Asia.

The net effect of these pressures was to suggest to the Japanese that the United States was contemplating unilateral revocation of the security bargain for East Asia, a bargain, furthermore, in which the United States was progressively less valuable to Japan in its role of military protector. Thus, if there was to be substantial Japanese rearmament, beyond the incremental growth that had already occurred and was itself quite substantial, one could not assume that it could be insulated from a greater Japanese political autonomy and assertiveness.

As in Europe, the dominance that has been exercised by the United States in East Asia and the Pacific has served to freeze many nascent conflicts in the region. A diminution of the American presence, especially if accompanied by a greater Japanese political (and perhaps military) role, would be likely to revive these underlying conflicts.

THIRD WORLD IMPLICATIONS

Much as familiar security relationships were changing in the areas of greatest concern to the United States in the postwar years, they were also changing in the Third World, where countries are increasingly less objects of superpower competition and increasingly more actors in their own right. Furthermore, the ability of the superpowers to project military power into Third World countries is increasingly suspect as Third World states acquire effective means of political organization and military resistance.

One should resist interpreting these trends as merely the product of recent years. They are, in fact, the culmination of historical developments that have been under way for decades. In retrospect the political and economic domination that the European states (and the United States) acquired over the peoples of Asia and Africa in the eighteenth and nineteenth centuries was the product of a fleeting military and organizational superiority. The European success was to carry with it the seeds of its own demise as the economic structures and technologies that had given rise to it were diffused to the subject peoples. The Europeans also transmitted to colonial peoples the national consciousness that had been an important component of the Europeans' ability to mobilize their technological resources for external conquest.

The first dramatic outgrowth of this process was the wave of decolonization that spread through Africa and Asia in the early postwar years. In the years since,

Third World states have demonstrated growing powers of resistance to attempts at superpower military intrusion. The inability of the United States to impose its will on Vietnam or Iran is matched by the frustration that the Soviets met in Afghanistan. The proliferation throughout the Third World of military technologies that put the military forces of even advanced states at risk, cheaply, and often without requiring great technical expertise, has added another element to the difficulties that advanced states have encountered in using their military forces to achieve goals in those parts of the world that formerly came under their colonial rule. To the proliferation of these relatively cheap conventional munitions must be added the proliferation of ballistic-missile technology, as well as capabilities relevant to the production of chemical and nuclear weapons.

A WORLD OF GREATER COMPLEXITY

Thus, a world of greater complexity could arguably be seen emerging in those regions where the United States has had important alliance and alignment relations in the postwar world, and in the Third World beyond. The most obvious manifestation of this complexity lies in a dramatic rise in the number of actors capable of setting their own priorities in international relations, as compared to the immediate postwar years.

Underlying this new complexity is a relative decline in the American economic position, and a consequent relative rise in the position of others. The early postwar years saw the United States in a position of overwhelming economic preponderance, a position that allowed us to encourage economic policies on the part of others that adversely affected our own immediate economic interests, as well as to maintain widespread military forces. Thus our economic position enabled us to cement security relationships with allies and friends by fostering allied economic development, as well as by offering more strictly military protection.

Decline in our relative economic position may itself be the function of a relative decline in the American capacity to innovate technologically. In previous decades new techniques and industrial processes tended to originate in the United States and then to spread to imitators elsewhere. Increasingly now, new developments and the economic advantages derived from them take place away from our shores and are imported here.

It should be observed, of course, that while American economic decline has occurred in a relative sense, as a share of the world economy, American output has continued to grow in absolute terms. Soviet decline has not only been greater in relative terms, it may well also have involved a shrinkage in absolute terms. Furthermore the principal beneficiaries of American decline are not the Soviet Union and its allies, but states which in the postwar years have generally been

American allies. Indeed, both superpowers are undergoing decline relative to certain other powers—most notably Japan, certain of the newly industrialized countries of the Third World, and some states in Western Europe (the latter more emphatically so if moves to further the economic unity of Western Europe bear fruit).

The effects of greater complexity are multiple, and likely to be contradictory. To some extent, greater complexity is likely to reduce the general level of tensions by making it less likely that disputes in discrete functional or geographical areas will reinforce one another. Given a larger number of actors autonomously setting their own priorities, it becomes more likely that countries in conflict over some particular issues may find themselves cooperating on others. Thus, two states in conflict over an economic issue may find themselves in need of each other's good will on a security question. For example, Japanese–American economic conflicts have historically been constrained by the felt need of the parties to avoid any action that might jeopardize their security relationship.

On the other hand, complexity may increase the potential for conflict by increasing the likelihood of miscalculation. In a bipolar world the attention of each actor is riveted on the other. Furthermore the two powers are in a position to adjust to each other's actions through their own respective internal actions. A superpower seeing another increasing its military potential, for example, can respond by increasing its own. The behavior of lesser states and the capabilities that those lesser states can supply to the superpowers are relatively unimportant compared to the behavior and capabilities of the superpowers themselves in governing the superpower relationship. Thus the actors in a bipolar relationship monitor one significant other closely and are less likely to suffer surprise. By making adjustments through internal means they are in a position to respond in a flexible and relatively precise manner to the actions of the other (although perceptual problems do not ensure that such adjustments will in fact always be responsive to the realities of external action).

In a world of many autonomous powers, each actor must monitor the behavior not of one significant other, but of many. This increased requirement for external intelligence suggests that the likelihood of surprise is greater. Furthermore in such a world the likelihood decreases that each power will be able to make adjustments to external actions by relying upon its own resources only. The importance of external adjustments, through international alignments and coalitions, will increase. The potential for changes in alignment introduces still another occasion for possible miscalculation or surprise.[1]

A world of greater complexity would have contradictory effects on American national security. Decline in Soviet military potential would ease tensions in those regions—Europe and Northeast Asia—concern for the security of which gave rise to American military response at the outset of the Cold War. The necessities of domestic renewal might well make the Soviet Union more dependent on international economic relations and thus more reluctant to offend actual and potential economic partners in order to secure some security advantage. But as such a decline results in the effective elimination of Soviet control over such areas as Central Europe, the potential would increase for a variety of conflicts arising suddenly through the interaction of the domestic and international behavior of the newly autonomous powers that will replace Soviet dominance. In East Asia a more activist Japan might indeed take upon itself some of the security responsibilities now borne by American forces, but at the cost of renewing regional animosities and fears.

A world of greater complexity might introduce numerous new threats to American national security as a variety of newly autonomous actors operate on the world stage. Furthermore, the very multiplicity of these newly autonomous actors may make it more difficult to predict the character, timing, or location of the threats that might arise. However, this very complexity may reduce the necessity for direct American planning and involvement in meeting those threats. These threats may engage others as well as the United States, and we may more often find ourselves blessed with the luxury of relying upon the self-interest of others to meet problems of potential concern to us. Thus, Vietnamese expansionism in Indochina has been contained by regional opposition, by Soviet reluctance to allow the continuation of activities complicating her relations with China and other powers, and by Vietnamese domestic exhaustion. American involvement in the process of containment has been marginal. This situation may well prove to be a model for others.

The above view, that the security environment for the United States will likely take on greater complexity in the future than it has had for much of the immediate postwar period, was challenged by those who questioned the likelihood of fundamental and lasting change in the Soviet position in international relations and in Soviet external behavior. If, in fact, the Soviet concentration on internal reform was merely a truce in the Cold War and not its end, we might well see a return to Soviet external aggressiveness at some point in the future. A powerful and vigorously hostile Soviet Union would remain the preeminent driver of American security and foreign policy that it has been since the early years of the Cold War. Certainly the notion of such a truce, a time to renew strength in order to later prosecute an intractably hostile relationship with capitalist powers, has good Leninist roots. Such was indeed the policy followed by Lenin in the early days following the Bolshevik Revolution. Lenin began by accepting the tremendous territorial losses of the Brest-Litovsk treaty of 1918 with the Germans in order to remove Soviet Russia from World War I. Lenin's purpose in accepting this settlement was to gain for Russia a respite in external affairs that would enable him to secure the survival of the new Soviet regime domestically. Later, Lenin was to adopt the New Economic Policy, under which some degree of capitalism was allowed to return to Soviet territory. Cooperation with the West was deliberately encouraged to acquire domestic investment and technology. The object of this policy was to gain the economic and technological infrastructure that would enable Soviet Russia to support a modern military machine. Still later, Stalin was to pursue popular-front strategies in the 1930s, designed to defuse Western antagonism to the Soviet regime. This would allow time for Stalin's forced industrialization to produce Soviet military power sufficient to resist Germany. Stalin's failure to find common ground with the Western powers in the period immediately preceding World War II was to lead to a pact with Germany, defining respective spheres of influence in Eastern Europe and encouraging Germany to deflect its aggressive intentions to the West after overwhelming Poland. This was yet another expedient to buy time until Soviet military power was sufficient.

Nor can we presume that the current amelioration of ideological hostility to the West now apparent in the Soviet Union is irreversible. The Soviet view of international relations has long sprung from the premise that these relations are rooted ultimately in class conflict, and that these conflicts are ultimately unbridgeable and must eventually be resolved by forcible, violent confrontation. The implications of this view have been successively modified in the postwar period, to suggest that world war, with the massive destruction that would follow in its train, is no longer inevitable as part of the historical process. Now the fundamental picture of class-based conflict in international relations itself is questioned. Those skeptical of the durability of this latest view suggest that ideological propositions have functioned for the Soviet government as convenient formulations designed to justify the policy of the moment. If return to an aggressive external posture is found necessary in the future, the skeptics would hold that a new basis in ideology could then be found to support such a posture.

One may doubt that ideological positions are as flexible as this view would suggest. In particular, one must doubt that the centrality of the class struggle, once dismissed, could easily be resurrected as the philosophic basis of foreign policy. In a more fundamental sense, one may doubt that the permutations of ideology, even if they could carry this sort of flexibility, can be, or have ever been, central in determining the importance of the Soviet Union for American policy. Ideological hostility to the West persisted in the Soviet Union for many decades after the October Revolution without making the Soviet Union the central threat to American security. Indeed, ideology did not prevent the Soviet–American alliance against Nazi Germany during World War II. What brought the Soviet Union to the center of American concerns was not its ideological posture, but profound changes in the structure of international relations which that war conspired to bring about. The war eliminated Germany and Japan as political, military, and economic actors on the world stage and enfeebled Britain and France. What later came to be called the Third World was largely chaotic. The Soviet Union came to occupy the center of American attentions largely because for a time there was no other object in view.

Unless some process occurs that reverses the growth of complexity in international relations, in the form of a rise in the number of actors capable of setting their own priorities and resisting superpower intrusion, it is difficult to see how the Soviet Union could come to occupy such a central position again from the American perspective, no matter what the evolution of Soviet ideology. This is not to suggest that the Soviet Union will cease to be regarded as a threat against which military preparations should be made in the United States. For the foreseeable future the Soviet Union is likely to loom as the single most important external concern for American policy. It is to suggest that the role of diplomatic maneuver in managing that concern may increase at the expense of the acquisition and deployment of military forces. It is also to suggest that the automatic necessity of direct American involvement in countering Soviet initiatives will decline.

Those who doubted the view that the security environment would be characterized by greater complexity could also question the likelihood that the trend toward the emergence of new decision-making centers in world politics would continue. They could and did argue that Japan's inherent vulnerabilities placed a cap on its ability to sustain an independent position in international politics, no matter how powerful its economy might appear. Geography and lack of domestic resources inevitably make the Japanese dependent on outsiders to a much greater degree than are the existing superpowers.

In Europe skeptics could argue that to dwell on the fissures that periodically appear in the alliance is to ignore its underlying continuity and the long-term common interests and institutional factors that will continue to bind it together. The overt threat from the Soviet Union may indeed decline. Geography dictates, however, that concern about Soviet military potential cannot be totally ignored in Western Europe. To be sure, a united Western Europe would have the demographic, industrial, and economic potential to be a military power superior to the Soviet Union and could thus provide for its own defense without need of the American alliance. But a moderating Soviet threat would remove this incentive for European political and military unification to follow the measures of economic integration now in train. Even an increase in the Soviet threat would hardly suffice to motivate the Europeans to surmount the historical, cultural, political, and psychological obstacles to unity. Unless such an increase was accompanied by complete American withdrawal from alliance commitments, the preferred solution of most Western Europeans to their security problems would remain as it has been since World War II. Alliance with the United States allows them to deter Soviet threats while at the same time avoiding the full costs of mounting such an effort on their own. Thus, increases in European political autonomy and economic power will remain constrained by the necessity to remain in alliance with the United States. This general conclusion of the skeptics was applied as much to West Germany—even to a united Germany—as to other Western European countries.

Nor did the skeptics accept a fundamental change in the direction of autonomy on the part of states in the Third World. Under this rubric lies a vast range of national experience, ranging from the newly industrialized countries, which have proven capable of advancing national wealth, to others mired in poverty. In many of these states governmental authority is weak. In many of them boundaries and political structures reflect, not indigenous ethnic divisions, but the arbitrary bureaucratic dictates of their former colonial masters. Weak domestic government and arbitrary or illegitimate boundaries are a recipe for continual strife, ranging from civil conflict to international war. In some cases conflict results not only in the wake of the departure of European dominance, but from the reemergence of long-standing and indigenous conflicts. The Iran–Iraq war provides an example of a conflict that has endured for hundreds of years, quiescent for long periods, but flaring up periodically when one or the other side thinks it sees an opportunity to gain advantage. With all these opportunities for potential conflict abounding, the possibilities are legion that indigenous powers will attempt to enlist outsiders in their support. Furthermore, skeptics would argue that recent cases of effective resistance by Third World states to outside intervention may be due merely to local topography or the potentially reversible state of military technology and should not be made into universal generalizations.

If indeed an environment of greater complexity, approaching multipolarity—in the sense of a number of actors possessing political autonomy in decision making to roughly equal degrees (even if their political, economic, and military capabilities are not equal)—does lie in the offing, such an environment, while encouraging or constraining certain policies, will not determine policy choices. Here, much will depend on the nature of Soviet external behavior in the coming years and, perhaps even more crucially, upon the nature of the American interpretation, both of Soviet behavior and of the wider world generally. For it is how Americans perceive external reality that will largely influence the policies which they will support or allow.

A CHANGING INTERNATIONAL SYSTEM

As the new American administration took up its task in early 1989, the view that world politics was on the verge of a qualitative change from the familiar patterns of four decades, while asserted widely enough, remained problematic. While trends hinting at radical change could be noted, they seemed balanced by those that seemed to presage continuity. The question remained open. The administration certainly felt confident it could base policy on the premise that familiar patterns in the international environment would remain, subject at most to slow and evolutionary change.

By the early weeks of 1990, however, the conclusion seemed unavoidable that—whatever form a new international order might take—the system that had prevailed since the end of World War II was no more. Questions that months before could legitimately be regarded as open and uncertain had been decisively answered. Factors suggesting change in the structure of the international system had predominated over those suggesting continuity. While two previous administrations that had proclaimed novelty were forced by events to acknowledge continuity, an administration that proclaimed its expectation of continuity found itself paradoxically forced to confront novelty.

The most dramatic developments related to the sudden decline in the Soviet political position in Europe and the correlative revival of a united Germany as a major actor in the European, and world, arena. The immediate genesis of these developments has perhaps been the greatest surprise of all. For they were catalyzed by actions of the Soviet Union itself. The Soviet government set in train, and indeed encouraged, developments that swiftly led to the dismantling of that political order in Europe, the maintenance of which for so many years it had equated with security itself.

By signaling that it would no longer support ideologically congenial regimes in Eastern Europe with the ultimate sanction of the Red Army, the Soviet Union ensured that these regimes—almost totally devoid of domestic legitimacy—would collapse in a cascading chain in a matter of weeks. The Soviet government may have hoped to stabilize the political situation around a two-track solution in which domestic change in Eastern Europe would coexist with continuity in the international orientation of the Eastern European governments. If so, doubts as to the stability of such a solution were borne out, as newly autonomous governments in the region began to demand the withdrawal of Soviet troops from their territory. To these demands, the Soviets had little choice but to acquiesce. These military withdrawals may not be incompatible with the Warsaw Pact in a strictly formalistic sense, but they cast doubt on the continued coherence of an Eastern European alliance in a politically meaningful sense, and they certainly end any utility that the Pact may once have had as a military grouping.

Once domestic autonomy was granted to the Eastern European clients generally, its application to East Germany in particular could not be prevented. And domestic autonomy in East Germany swiftly took the form of immense pressure to unify the German states politically and economically. After some initial hesitation, the Soviet Union was to give way to this pressure by removing its opposition to a united Germany.

It is extraneous to our purposes here to speculate upon Soviet motives in allowing, and indeed encouraging, such dramatic events. No doubt the Soviet leadership was merely drawing the correct political conclusions from its recognition of a long-term economic decline that rendered the costs of maintaining the old political order in Europe increasingly burdensome. If so, what must be striking is the willingness of the Soviet government to draw such sweeping political conclusions and swiftly to act upon them. For the conventional wisdom is that governments tend to make only the minimum changes in policy absolutely necessary to respond to specific problems and will not voluntarily abdicate positions of advantage except when compelled to do so by immediate and overwhelming dangers.

Whatever the explanation of the *domestic* motives for Soviet behavior, the *external* implications of that behavior are immense, and effectively irreversible. At a stroke the Soviet Union has vastly reduced its political influence in the center of Europe and made immensely more difficult any future attempt to bring the Red Army to bear against Western or even Central Europe. Whether or not current Soviet policies are merely tactical retreats to be followed by future reversion to aggressive policies has become a largely irrelevant question. For such a reversion to a more aggressive policy would represent a much reduced potential threat to Western Europe because of the constraints introduced by these changes in the structure of power in Europe. Furthermore, if this Soviet withdrawal from the former political position is indeed dictated by the Soviets' economic decline, their capacity to mount such a future military threat to the West will become progressively attenuated, whatever the nature of the external constraints placed upon the Soviet Union, unless one can foresee at some future period a profound reversal of its economic fortunes.

As noted, Soviet political retreat from Eastern Europe carried with it an inability to preserve the division of Germany that emerged from World War II. Resolution of the German question thus became a matter for decision by the Germans themselves. The result has been the emergence of a united Germany that will be autonomous in her foreign and security policy.

CHANGES IN AMERICAN FOREIGN POLICY

These two considerations—the Soviet withdrawal from dominance over Central and Eastern Europe and the emergence of a unified Germany—must eventuate in a radical change in the nature of American foreign policy commitments in Europe, and in the alliance and security policies that have supported our policies. Because the nature of our commitments in Europe have shaped our policies elsewhere, we can expect a general impact on our policies throughout the globe.

For forty years our external outlook has been shaped by the conviction that the Soviet Union represented a real, and imminent, military threat to overrun the European continent, or to cow Western European countries into submission to Soviet wishes out of fear that such a threat might be made reality. To meet this threat we felt compelled not only to declare our willingness to meet it should the need arise but to permanently station military forces on the continent. These convictions were shared—were indeed largely induced—by the parallel conviction of governments in Western Europe that only the presence of such a permanent American military force could reliably prevent the Soviets from capitalizing on their threats.

The result of these parallel convictions was an alliance system in which the interests of all parties coalesced upon a common conception of threat from the Soviet Union. This coalescence of interest was supplemented by agreement on an integrated military structure as the principal means to give effect to this common interest.(While France was to withdraw from formal participation in this integrated structure, it has never demanded the abolition of the structure and has cooperated with it throughout.) To say that some of the parties to the alliance may have seen it as serving other interests—notably as restraining Germany—is not to deny the overriding importance of the perception of threat from the Soviet Union in giving NATO the cement that held it together.

It is unlikely in the extreme that these commonly accepted conclusions as to the threat—and the means to meet that threat—which have sustained NATO for so many years can be maintained. If universal perception by the allies of an imminent military threat from the Soviet Union declines, then we can expect the integrated military structure to decline in importance and the alliance itself to lose its coherence. The alliance may well continue in existence juridically, reverting in the process to its original form as a treaty of guarantee. But it will become less and less influential as a vehicle around which the policies of its members coalesce.

If NATO does indeed decline in importance in this sense, what sort of political structures may we expect will take its place? What requirements will these new structures lay upon American policy?

Perhaps the central question of European political structure is the degree to which it will be dominated by Germany. Will Germany exercise an increasingly autonomous influence over the Central and Eastern European regions? Or will Germany remain tied to a more Atlanticist posture through continued institutional relationships of one sort or other with France, Britain, and the other states of Western Europe?

German ties to Western Europe may be fostered through a variety of existing institutions, such as the European Community, the Council of Europe, or the Western European Union. Such relationships would carry with them the implication of continued military cooperation, at both the conventional and the nuclear level; they would also serve to reassure the Soviet Union—and some of its erstwhile clients—by placing a measure of restraint on German policy toward the East.

While the Conference on Security and Cooperation in Europe (CSCE) is put forward by many—including the Soviet government—as a substitute for the present structure of alliances in Europe, it seems doubtful that such a broad grouping of thirty-five states could exercise the restraint on Germany that continued institutional relationships with Western Europe would provide. The CSCE will at best serve as a forum to legitimize arrangements reached through other mechanisms.

Germany without such relationships would be an object of greater concern to neighbors in the East and the West. The Soviet Union would no doubt attempt to exploit such concerns diplomatically by attempting to form political combinations of states that believed themselves threatened by Germany. The result might well be a complex pattern of shifting alignments, a pattern that might offer innumerable occasions for conflict through miscalculation.

Which of these political constellations comes to pass will largely be determined by European decisions in which the American role will not be a great one. In neither of them, however, does a permanent American presence seem required, or indeed sustainable. It seems unlikely for the foreseeable future that the Soviet Union will be capable of presenting a conventional military threat to Western Europe. Insofar as the Soviet Union remains a preeminent nuclear power, the threat that these weapons represent can be dealt with by American forces based elsewhere than in Europe.

Such a diminution of a Soviet military threat to Europe will allow for the possibility of major reductions in American nuclear forces—at both the strategic and the lower levels—and a major reorientation of their missions. The size and character of American nuclear forces have been driven largely by the perceived need to extend the nuclear deterrent to cover major attacks against certain of our allies, and by the perceived need to attack large numbers of targets to make such a deterrent threat credible. Changes in the political order in Europe may well eliminate the perceived necessity to extend the nuclear deterrent in this fashion. They may also reduce considerably the number of targets thought necessary to retain deterrent credibility against the residual threat. Furthermore, military targets less difficult to destroy than the strategic forces of the opponent may well come to be thought sufficient. The extent of reductions in nuclear forces will be guided not only by changes in political relationships and the strategic implications that might be deduced from them, but also by the arms-control process itself. Here, demanding technical requirements to ensure that the opponent is observing agreements and not deriving asymmetical advantage from them will exert a powerful effect. Additionally, we should note that major reductions in forces will have the effect of making the forces of smaller nuclear powers—Britain, France, China, and perhaps others—of greater military significance relative to superpower arsenals. There are limits beyond which the superpowers will consider reductions in their own forces without bringing the smaller powers into the negotiating process. Needless to say, a multilateral nuclear arms-control process may prove even more cumbersome than the present bilateral process.

If major reductions are in train, arguments for the addition of strategic defenses might gain validity. Such defenses would be more effective against major attacks launched from reduced strategic arsenals. In addition, they would have great utility in guarding against the residual possibility of accidental small attacks by a major power, by minor nuclear powers, or by nonstate actors. It is unlikely, however, that deployments of such defenses will be determined by such arguments. Far more powerful in such a determination will be domestic political attitudes. While localized defense against attacks may prove acceptable, area defense systems against major attacks may not be acceptable. It seems improbable for the foreseeable future that such attitudes would allow the deployment of such a new category of weapons, barring some catastrophic event.

The United States may find it prudent from time to time to give assurances of support to one or another European grouping. Indeed, Europeans may from time to time require such reassurance. Insofar as these assurances might require a military component for credibility, this requirement can also largely be satisfied by forces mainly based in the United States, supplemented in this case by a capability to deploy to pre-prepared sites in Europe on short notice, and by a capacity to mobilize large forces on longer notice should major threats reemerge. American conventional forces designed for the residual contingency of a major conflict might well revert to the character that they took during the interwar period: a small permanent cadre capable of expanding to a major force on notice. What is lacking is any demonstrable necessity for the continued maintenance of major American deployments on the Continent. While American continental commitments may remain, they will likely be ad hoc and short term rather than permanent, and conveyed primarily by political and diplomatic means rather than by the physical presence of military forces.

In contrast to the situation regarding nuclear arms, where it is likely that the arms-control process itself will continue to powerfully shape the size and structure of forces, it seems probable that domestic and international political influences will predominate with regard to the evolution of conventional forces. Indeed, force limitations proposed by political leaderships have already outpaced the negotiating process designed to operationalize them. We should expect this process to continue. Arms control in its traditional form of attempting to specify precise quantitative and qualitative limits may serve only to place floors underneath which it may be regarded as imprudent, for political reasons, for forces to sink. It is an arguable proposition that arms control in this form should be abandoned altogether and replaced by a process that emphasizes confidence-building measures.

The dramatic events in Europe have had no parallel in other environments of relevance to American security. In these arenas we witness instead the incremental evolution of long-developing trends toward complexity already discussed. In East Asia and in the Third World we are likely to see a continued decline in

Soviet influence and involvement and the increasing dominance of local conflicts, disputes, and concerns in the politics of these regions.

This is not to suggest that American interests can in no way be threatened by an increasingly complex world. It is rather to suggest that both the character of the remaining threats and the capacity of the United States to meet those threats with military means will be limited in ways we have not experienced in many decades.

We may well continue to experience threats to our ideological well-being or our economic interests, as well as myriad localized threats to our citizens and physical assets. But these threats are unlikely for the foreseeable future to emanate from powers capable of militarily dominating the Eurasian landmass. In the absence of such threats, American domestic opinion cannot be expected to sustain defense expenditures at Cold War levels.

The international environment into which we are entering will in no sense be characterized by equality among the powers in it. There will remain vast differences in economic and military potential. But the diffusion of advanced military technology in a variety of forms will tend to render direct military attack increasingly costly to an attacker. The recent war against Iraq, while seemingly contrary to this trend, may, in retrospect, only underline it. For swift, decisive, and relatively cheap victory was achieved largely because of a substantial technological advantage that will be a vanishing asset. This suggests that direct military interventions—and the forces capable of mounting them through the sustained occupation of territory against forcible opposition—will grow increasingly less attractive to the United States, except on the smallest of scales. We will, nevertheless, wish to exercise influence from time to time, and to be in a position to offer local governments support and alignment. This suggests an increasing emphasis on forces designed to show presence, to punish at a distance, or to offer a local ally militarily relevant support without direct involvement. Such forces will increasingly be based at home rather than be dependent on increasingly problematic overseas basing.

Indeed, both in Europe and in Asia—and perhaps in some regions elsewhere—American security policy may increasingly not be focused on direct threats to our interests at all. Our continued presence in certain areas may be desired by many regional parties as a device to prevent or retard the sharpening of regional animosities that our departure might bring. Any such presence, however, is likely to be largely offshore and symbolic in character and will not involve major forward-deployed land contingents. The extent to which the United States responds to such desires for continued presence in regions in which our previous principal motivation for such a presence—Soviet involvement—has declined, as well as the duration of such presence, will largely be a function of the evolution of American domestic opinion.

An increasingly complex international environment no longer structured primarily by the Soviet–American relationship is unlikely to be an especially harmonious international environment. The swift reemergence in Europe of international and interethnic conflicts buried since the 1930s provides ample illustration that it will not be so. The spread of advanced forms of military technology suggests the potential for great destructiveness should conflict take the form of major violence. Many of these potential regional conflicts may well be constrained, however, by the difficulty that local powers will perceive in deriving a demonstrable advantage from unleashing them. American diplomatic efforts and technical assistance may well be useful at the margin in increasing such constraints. A few such conflicts may still threaten American interests directly; whether major American military engagement will prove efficacious or feasible in preserving those interests more than momentarily, in the absence of regional, political, and economic factors serving simultaneously to support such interests, remains to be seen.

Nor can we take much comfort from the assertion that economic processes will overtake politics in importance. Economic interchange is as likely to lead to protectionism and beggar-thy-neighbor policies as it is to lead to the spread of mutual benefit. This seems especially ominous in an environment in which the United States has declined from its postwar dominance of the international economy and no potential successor seems capable of taking over single-handedly. Indeed, the next decades may well establish for us whether economic transactions themselves—in the absence of a single power exercising a hegemonic position in both the political and the economic spheres—are capable of ameliorating conflict.

Unless some revolution occurs in military technology, however, large-scale warfare will remain an unattractive instrument for the resolution of those conflicts that will occur. Only in those limited places where great—and probably fleeting—qualitative technological imbalance exists will war be an attractive option. It will remain especially unattractive for a United States that will continue to retain many of the advantages that its historic insularity has given it. We will doubtless continue to remain active in preserving our interests and ameliorating potential conflict in a variety of regions of the world. When we choose political instruments to further these policy goals, however, it is increasingly likely that these means will emphasize a capacity for flexibility of alignment and position. When we choose military instruments, we will be increasingly drawn to those that can be exercised at a distance and with minimal risk of loss or entanglement.

NOTES AND REFERENCES

1. These arguments reflect those developed by Kenneth N. Waltz in "The Stability of a Bipolar World," *Daedalus* 93 (1964).

2

Some Considerations on the Soviet–American Relationship in the 1990s

Robert W. Tucker

I

In considering current speculation on the prospects for the Soviet–American relationship in the coming decade, it may be useful to recall past speculation on this relationship. The record is not one to inspire a great deal of confidence in present efforts to foresee the future. At the close of the 1940s the best known projection of the Soviet–American relationship was, of course, the National Security Council document NSC-68. That now-famous document, written in the shadow of the Chinese Communist victory and the Soviet Union's development of the atomic bomb, forecast a year of maximum danger that would be reached in 1954. The 1950s ended with what was seen as an ominous disparity in missile development and an almost equally ominous disparity in rates of economic growth, prompting widespread fears that the terms of the balance of power were shifting sharply to our disadvantage. A decade later the domestic trauma of Vietnam led many to conclude that the 1970s would find a general American withdrawal in the world, and certainly in the developing world. When the 1970s drew to a close the Soviet–American relationship was seen to have entered a period even more dangerous perhaps than the period marking the early years of

---

Robert W. Tucker • The Paul H. Nitze School of Advanced International Studies, The Johns Hopkins University, Washington, D.C. 20036.

the Cold War. In Europe and in the Persian Gulf the United States was considered to face a decade of peril during which it would have to mount its best efforts in order to reverse a balance of military power, strategic and conventional, that now favored the Soviet Union.

Today the prevailing outlook is radically different from the outlook that marked the earlier decades of the post-World War II era. Optimism rather than pessimism is the order of the day. The great conflict that only yesterday seemed so intractable and threatening has not only ameliorated, in the view of a growing number of observers it has all but run its course. A successful end of the Cold War is increasingly taken not as a matter of speculation but as a reality. Accordingly, the time has presumably come to think seriously about the kind of world we may expect to find beyond the Cold War. A vision of this world has already taken form. It is one in which war has become a thing of the past among the states of the developed world. Among the states of the developing world, war is also expected to decline markedly in significance as the process of development, increasingly determined by the values and institutions of liberal democracy, works its inexorable ways.

The prospect of a world without war is not, of course, a novel one. Few themes have been more persistent in the past generation than that of the growing disutility of force. But so long as the Cold War persisted, let alone appeared to intensify, as it did for a time, this theme could enjoy only limited favor. If the obsolescence-of-war view enjoys greater vogue today than ever, it is largely because recent and unexpected developments have given it a new persuasiveness. It is not the occasions on which force was actively employed in the 1980s that principally contributed to this result, because these instances do not clearly support any firm and sweeping conclusions. Instead, the markedly greater favor with which that view is received today is very largely the result of the visible failure of communism everywhere, though particularly in the Soviet Union, and the corresponding success of liberal democracy.

In light of the record of yesterday's prophets, may we place much confidence in today's seers? The latter increasingly look forward to the prospect of an accommodation with the Soviet Union that will put a definite end to the Cold War. Those who remain skeptical of this prospect, however, must necessarily point to the past. We have already experienced a number of occasions, they argue, on which the Cold War appeared to be receding into history only to find that the appearance was deceptive. If this past is any guide to the present, must we not expect a similar outcome? This is not to say that the great conflict will remain unaffected by developments that have given rise to such widespread hopes and expectations of a new world. Clearly, it has already been affected by these developments and may well be even more affected in the decade to come. But it is one thing to view developments affecting a conflict that nevertheless persists, though perhaps in markedly moderated form, and quite another to view the same developments as transforming the conflict altogether because they have altered, and irrevocably so, the essential sources and conditions of conflict.

It is the latter view that is commonly taken by those who find that we have come to an end of the Cold War, if only we have the wit to see this. The past can have but limited relevance to the present and future for the simple reason that we have presumably come to a turning point in history where what has gone before has only a limited bearing on what is to come. This historical discontinuity presumably holds true even with respect to the immediate past. Only on this basis can the first half of the 1980s now appear quite irrelevant. For it was, after all, only yesterday that we were still living in a world that was quite traditional in its politics.

II

The debate that has arisen over the prospects for a new relationship with the Soviet Union is, in the main, one between those who believe that a change of fundamental significance has taken place in the Soviet Union's relations with the Western world and those who believe that these relations, though they have for the time being markedly ameliorated, remain essentially unchanged. The evidence put forth by the two sides is by this time familiar enough and may be summarized. To support the view that Moscow has changed and profoundly so, the proponents of a transformed Soviet foreign policy point in the first instance to the words of Soviet foreign-policy elites. These words are indeed strikingly different from the words of the past. In place of the struggle between two social systems, capitalism and socialism, we now have the interdependence of humankind throughout the world. Instead of finding in peaceful coexistence a specific form of class struggle, it is now given a literal interpretation and equated with the universal desire for peace and security. Rather than finding in military power the great and reliable means of national security, the danger of excessive reliance on military power is emphasized and the growing disutility of force is stressed. Nor is one's security seen as something that is to be obtained at the expense of others. The mutuality of security is emphasized as is the modesty of means—reasonable sufficiency—by which this goal is to be realized. While some of these themes are not novel, having been given occasional expression in the past, many of them clearly are. In their totality they express a transformed outlook.

If taken at face value, and as indicative of future actions, the new thinking must be considered tantamount to a revolution in Soviet foreign policy. Those who find something akin to a revolution indeed occurring also point to acts that have accompanied words and that are seen to dictate, in their cumulative impact, the same conclusion. The Soviet withdrawal of military forces from Afghanistan together with the cooperation that Moscow has shown in moving toward the settlement of several regional conflicts are part of this new pattern of action. So, too, the willingness to accept a degree of intrusive inspection and verification of arms-control agreements that Soviet governments had always rejected in the past. In concluding the 1987 agreement on banning intermediate-range nuclear forces in Europe, the Soviets not only acquiesced in measures of inspection and verification that they had before resisted, they also agreed to asymmetrical cuts in nuclear forces, something that they had also resisted in the past.

The acceptance of the principle of asymmetry was further manifested by Moscow in its proposals on the reduction of conventional forces in Europe. In the wake of a unilateral decision announced in December 1988 to reduce Soviet military forces by 500,000 troops and to withdraw over 40 percent of Soviet tank divisions from Eastern Europe, Moscow proposed in early March 1989 to reduce its conventional forces in Europe by more than half in return for negligible North Atlantic Treaty Organization (NATO) cuts, provided that the reductions are made not only in battlefield arms but in aircraft and troops as well. A conventional forces treaty was signed in 1990.

Along with these reductions of conventional forces, the Soviet government announced plans to reduce its overall defense spending 14 percent by 1991 while cutting its production of arms by almost 20 percent. These reductions had been seen by many observers as evidence of an intent to restructure the deployment and strategy of the Warsaw Pact in support only of defensive aims. Even that objective now seems beyond the reach of realistic policy, given the developments that have occurred since the fall of 1989. The collapse of Communist regimes in Eastern and Central Europe has now led to the demise of the Warsaw Pact as a military alliance. To date, the Soviet government has given little sign that it is prepared to take drastic measures to shore up its disintegrating alliance position.

The Soviet Union has not limited its diplomatic efforts to improve relations to the United States and Western Europe. The Soviet government has also undertaken to improve relations with its great Asian neighbor, China. Not only has it progressed toward the settlement of a number of issues that have marred Sino-Soviet relations in the past, but Moscow announced in early 1989 that it would soon begin the withdrawal of the greater part of its forces in Mongolia on the Chinese border. The withdrawal has begun.

In yet another area of vital interest to the Soviet Union, the Middle East, the same pattern is apparent. Moscow no longer follows a diplomacy that seeks to counter American influence in the region by tying itself closely to the radical regimes in Syria and Libya. Instead, it has undertaken to rebuild its relations with the moderate Arab regimes and it has taken a number of steps to establish a more normal relationship with Israel. Finally, the Soviets have cooperated, although at times grudgingly, with American efforts in the Persian Gulf.

Finally, there is the changed Soviet attitude toward international organizations that in the past Moscow has either shunned altogether or denied support to even though enjoying membership. Since 1987 the Soviet Union has voiced a desire to become a member of those organizations that heretofore it has identified exclusively with the economic and financial interests of the capitalist West. At the same time it has voiced support for a number of United Nations activities, including those in the field of peacekeeping, that it has opposed in the past and has begun to pay assessments that it had previously announced as unlawful. Even the World Court, an institution for which the Soviet government almost never had a kind word, let alone employed, has been subject to a new appraisal by virtue of Mikhail Gorbachev's 1987 proposal for strengthening the court through securing agreements among the great powers with permanent seats on the Security Council, agreements which would give the court jurisdiction to settle disputes arising among them in clearly defined areas. In 1989 the Soviet Union and the United States agreed to accept the jurisdiction of the International Court of Justice in disputes arising over several treaties dealing with terrorism and traffic in drugs. A new internationalism has come to characterize Moscow's outlook.

This, in brief, is the gist of the case made for the view that Soviet foreign policy has undergone profound change. The skeptics about change, and certainly of profound change, on the other hand, do not deny that words may be significant. They do insist that in matters of state, it is acts that are above all significant. Yet it is in reviewing recent Soviet actions that the skeptics remain if not unimpressed then at least unpersuaded that a fundamental change in Soviet foreign policy has occurred. On the contrary, the net balance of Moscow's recent moves are seen to have advanced its diplomatic fortunes, and most strikingly in Western Europe. For if it is true that the fortunes of the Warsaw Pact have markedly declined, it is also true that the prospects of the Atlantic alliance must also be expected to decline. Even the continued presence of American forces may no longer be readily assumed. With the denuclearization of Germany, a process that has already gathered considerable momentum, that presence may no longer be feasible. In the meantime, however, the Soviet Union continues today to hold out a military threat to Western Europe. That threat may eventually diminish if Moscow's promises of unilateral cuts in conventional forces are completed, if its apparent acceptance of the principle of asymmetry is once implemented, and if a restructuring of its offensive forces is followed through as promised. The reality, though, is that the Soviet government has markedly improved its diplomatic position on the basis of promissory notes that may never be met.

Nor is the present Soviet disposition to alter previous policies and to withdraw from previous commitments in the Third World without ambiguity. In a number of instances, diplomatic change may prove quite compatible with continuity of interest. In Afghanistan the withdrawal of Soviet forces has yet to result in the abandonment of the government in Kabul that Moscow installed. While calling for the mutual cessation of military aid to the two sides in the Afghan war, and for negotiations that would lead to a compromise settlement of the conflict, Moscow has continued its military support to Kabul and has not spared effort in doing so. Toward Iran, to take another example, the policy pursued by Gorbachev might just as well have been pursued by Gorbachev's predecessors. It is the case that Moscow no longer identifies itself with the policies of the radical regimes in Syria and Libya. At the same time it has been careful not to break openly with those regimes. The result of Gorbachev's new policy in the Middle East may only mean that the Soviet Union will henceforth enjoy the best of both worlds. While cultivating a new image of moderation, and rebuilding its relations with moderate pro-Western Arab states, the Soviet Union may succeed in obtaining a major role in the area, one that had previously been denied it. At the same time it may retain much of its influence with the radical regimes, an influence that might always be put to active use should circumstances necessitate a reversal of policy.

Then, too, there is the consideration that despite Gorbachev's insistent disavowal of force as an instrument of policy and despite his apparent conviction that military power has suffered a marked decline in utility, Soviet military production remains, by the Soviet president's own admission, virtually unchanged. It does so despite the state of near-crisis conditions that the Soviet economy has reached. *Glasnost* notwithstanding, the real extent of Soviet military expenditures continues to remain secret (and this despite Gorbachev's announcement in the summer of 1989 that the Soviet government is spending some 77 billion rubles per year for defense), a fact that itself must qualify confidence in how much Moscow has changed. This matter apart, why does the Soviet government persist in maintaining a level of military expenditure that, in proportion to the gross national product, is probably more than three times the American level? And why does it decline to exchange basic information on forces and force structure with this country, despite an earlier agreement in principle to do so? Until these questions are answered with a persuasiveness that they have yet to receive, skepticism must persist on how much the bear has indeed changed.

III

Significant policy consequences have been drawn from these respective positions. Those who believe that Soviet relations with the West are undergoing fundamental change by and large support a policy of doing all we reasonably can to support the government effecting this change. We must help Gorbachev in his efforts at home and abroad, it is argued, lest he fail and his conservative opponents come to power. Implicit in this argument is the assumption that our help may well make the difference between the success and the failure of Gorbachev's policies. Also implicit is the assumption that the failure of Gorbachev's policies at home and abroad are reversible and that his conservative opponents would presumably reverse them. Whereas the first assumption exaggerates the significance of the help we might give the Soviet government in its efforts to restructure the Soviet economy, the second assumption goes far toward accepting the skeptics' view that Soviet political liberalization, and what are considered the foreign-policy consequences, remain quite reversible (and not only by Gorbachev's opponents but even perhaps by a subsequently frustrated and disillusioned Gorbachev). But if this is the case and if it is true that our economic help cannot determine the success or failure of *perestroika*, it would seem to follow that we should be quite wary about the help we give Gorbachev. Although this help cannot ensure his success, his successors (or even a later Gorbachev) might well use it for purposes that we do not approve.

The difficulty in the skeptics' position is in determining what, if anything, might satisfy their doubts. Dismissing words (new thinking) and largely discounting acts, the skeptic makes demands for a measure of proof that are unlikely to be met. Thus the posing of litmus tests which, if met, are simply replaced by new tests. At one time the Soviet military withdrawal from Afghanistan was seen by many skeptics as a true test of significant change in Soviet foreign policy. This test was no sooner met than a new one was devised: Soviet abstention in Central America. Alternatively, a year ago the prospect that the Soviet government would soon undertake to make the dramatic changes it did in Eastern and Central Europe would have seemed remote. Yet the appearance of that prospect has not prompted many skeptics to revise their view that Soviet policy toward Western Europe has changed in any essential respect.

That the many changes Gorbachev has introduced in Soviet foreign and defense policies are all reversible is, in a literal sense, true. Some of these changes could not be readily or easily reversed, but they are all literally reversible. If the power and the will to employ it are there, territory vacated by armies may always be reoccupied. Spheres of influence in which a once-rigid hold has been loosened, if not simply abandoned, may again become spheres of influence. Given sufficient time, force structures that have once been altered may be changed back to their original disposition. And, of course, words may always be disavowed either by insisting that they were misinterpreted or by claiming that they were intended to apply to circumstances which are no longer relevant.

The reversibility argument has less to do with the nature of the changes to which it is applied than with the nature of the party promoting change. If changes in Soviet foreign policy are considered readily reversible, it is largely because these changes are undertaken by a government that is deemed to remain essentially unchanged. It is, then, the transformation of the Soviet state that would presumably provide assurance to skeptics that changes in foreign policy could not be readily reversed. But if this is so, it follows that the Soviet Union must undergo fundamental change in its domestic political structure before changes in foreign policy can achieve a credibility that they do not now have.

This is a very large order. But even if it were to be satisfied, why should it be assumed that changes in Soviet foreign policy would then be credible in a way that they are not today unless it is further assumed that this policy is first and foremost a function of domestic political structure (and of an ideology that plainly is of ever-decreasing significance)? So long as there is no significant devolution of power in the Soviet Union, the argument runs, Moscow will always be able to make sudden and substantial, even radical, changes in foreign policy, something that Western democratic governments cannot do.

This view that domestic political structure is the great determinant of foreign policy is not distinctive to the skeptic's position. Essentially the same view is held by those who believe that Soviet foreign policy has undergone a fundamental change. What separates the believer from the skeptic is rather that the former considers there has already been significant domestic political change in the Soviet Union, change that will be very difficult and costly to reverse, whereas the latter does not. This is a critical, though often unarticulated, difference and not what is often seen as such, that is, a difference in assessing motivation. Here there has instead been from the outset a substantial measure of agreement. That Gorbachev and his associates have taken the course they have is generally considered the result of failure at home and abroad. The Soviet economy, after decades of neglect and mistreatment, is in very serious straits. Unless it undergoes far-reaching reforms, the prospect is almost certain that the Soviet Union will decline precipitously in status. In addition, Soviet foreign policy over the past fifteen years has succeeded in creating a global coalition arrayed against it. Rather than having successfully frightened the world into making concessions to Moscow's sway, the seemingly endless accretion of military power only served to provoke an unexpected resistance. There is general agreement that the words and acts of the Soviet government today are motivated above all by this twin crisis and the need to overcome it.

It is apparent that Gorbachev's words and acts are subordinated to the necessities imposed by the crises in the domestic economy and in foreign policy. Will these necessities override intention and, in the end, even make intention irrelevant? If they may be expected to do so, it does not much matter what Gorbachev's initial intentions may have been. This is the argument of those who see the Cold War as having come to an end. They argue that even if Gorbachev's intentions are not markedly different from those of his predecessors (a difference that many, nevertheless, clearly credit), this will not matter a great deal in the end. For the necessities imposed by the course he has begun at home will dictate a foreign policy of moderation and will do so for at least a decade, if not longer. (And some have taken the position that even if he were to lose power, his successors would perforce follow a similar foreign policy, for they would find themselves governed by the same constraints.) By contrast, skeptics question whether the end result of the Gorbachev domestic reforms will entail radical domestic political change. In the absence of such change, they find little assurance that a foreign policy of moderation will be followed.

There is no way of satisfactorily settling this difference, which turns on the nature of and prospects for fundamental change in the domestic political structure. That one side views the prospects for a real devolution of power optimistically and the other side pessimistically is less significant perhaps than the fact that both equate the future of Soviet foreign policy, and thus the future of the Cold War, with domestic determinants. Although ideology has waned to a point few would have expected at the outset of the Cold War, the view that prevailed then—that Soviet foreign policy is the outcome of domestic factors—continues to prevail today.

IV

How far may the moderation of the Soviet-American relationship be expected to go in the decade ahead? In the Third World the response commonly given by observers is that we may expect substantial, even dramatic, change for the better. The reason for this expectation is that the Soviet leadership no longer entertains the outlook toward the developing countries that it once did. It no longer does so because it has come to appreciate that while its former policies in the Third World were a major factor in poisoning the Soviet Union's relations with the United States, these policies did little to compensate by way of enhancing Soviet position and influence. Even when success did seem to attend Soviet policy, more often than not it proved to be evanescent in nature. Moreover, the costs of empire as well were increasingly disproportionate to its benefits. In a period of economic stringency this consideration alone is found likely to place severe limits on the Soviet role in the Third World.

These considerations seem, on balance, persuasive. There does appear to be a genuine change in Soviet attitudes toward the kind of role they are now prepared to play in the developing countries. This attitude contrasts strikingly with the outlook entertained in the mid 1970s, when the Brezhnev regime was prepared to jeopardize its relationship with the United States rather than to abandon its aspirations in the Third World. It was these aspirations that, when given active expression, did in fact do more than anything else in turning American opinion against the détente concluded by the Nixon administration.

Is it reasonable to expect, then, that the 1990s will find a Soviet Union that in the light of its experience and as a result of economic necessity has by and large withdrawn from the developing world? A policy of general passivity, if not of complete abnegation, is in all likelihood too much to expect. There are, after all, a number of things that the Soviets may still do in the developing world that need not prove particularly burdensome economically—arms sales, which give every sign of being continued, may prove quite the opposite—and yet yield a measure of influence. It would be surprising if Moscow abstained from such activities, provided only that the United States does not make their cessation a necessary condition for a relationship of détente with the Soviet Union. To what extent the American administrations will seek to do just this is difficult to surmise, although the Bush administration in its first months in office gave more than one indication that it was disposed to use a weakened and vulnerable Soviet position as an anvil on which to try to hammer out desirable settlements of Third World conflicts. But how far and with what success this strategy may be pushed is difficult to estimate. In part, it will also depend on the larger relationship with the Soviet Union.

At the center of that larger relationship is, of course, the issue of a European settlement that would consolidate the dramatic changes that have occurred since 1989 in Eastern and Central Europe while providing for a security arrangement that would give both the West and the East such assurances as may be necessary for stability.

It has already been noted that Soviet foreign-policy experts, official and unofficial, have with increasing frequency speculated privately that the Soviet Union will have withdrawn militarily from Eastern Europe by the close of the 1990s. That speculation is, moreover, only the logical consequence of the new thinking in Moscow, with its emphasis on the self-determination of the Eastern European states to pursue whatever course of political and social development they may wish. What if that course should take a state increasingly toward a market economy as well as toward political institutions that are also increasingly liberal? In this event, it has been argued, the Soviets would prove accommodating so long as the state in question continued formally to designate itself as socialist and to remain politically in the Warsaw Pact. This suggests that Moscow has already accepted in principle the solution urged by many in the West, that security be henceforth defined in military terms alone rather than in both military and ideological terms.

If the Soviet Union has now changed its long-held position, this is clearly a development of the greatest importance. It may turn out, however, that the change will again be reversed. For the question persists: Why should Moscow be content with Eastern European states that, despite their internal evolution, continue to call themselves socialist, if this is increasingly little more than a bow to form and is devoid of substance? One possible answer, of course, is that the Soviet government now has a very different view of its security requirements. But the evidence here remains inconclusive.

Then, too, there is the second condition of a satisfactory Eastern European settlement: that these states remain politically in the Warsaw Pact. Does remaining in the Warsaw Pact imply, in turn, the obligation of member states to continue to allow the garrisoning of Soviet forces? Even if a settlement were to stipulate the right to garrison Soviet forces, it is unreasonable to expect that the retention of such forces would not be subject to the consent of the receiving state. How long would a Poland or Hungary abide by this agreement to entertain Soviet forces when they have governments that are no longer controlled by Communist parties obedient to the Soviet Union and when their economic salvation depends upon the development of ever more intimate relations with Western Europe?

If the Soviet Union were no longer to keep forces garrisoned in Eastern Europe, or if garrisons there might nevertheless be removed at the insistence of the receiving state, the Warsaw Pact would thereby be transformed. It might well become—if it has not already—a paper alliance, a mere form without substance. Would the Soviet Union be satisfied with this, particularly if the retraction of its military power, together with the erosion of its residual control over Eastern Europe, is not paralleled by either the withdrawal or the substantial reduction of the American military presence in Western Europe? Of course, if the retraction of American military power is assumed to proceed concomitantly with the retraction of Soviet power, that is another matter entirely, for in this event the Soviet withdrawal from Eastern Europe would be compensated for by the removal of American power. In the absence of American power, and in the absence as well of Western European political unity, the Soviet Union might well afford to do away with its military presence in Eastern Europe, because its geographical proximity and weight would confer on it a position of predominance not only in Eastern Europe but quite likely in Western Europe as well. Even an American presence that is much weakened might possibly prove a satisfactory trade-off for a Soviet withdrawal from Eastern Europe. Without these prospects, however, doubt must persist over what would prompt Moscow to withdraw its forces altogether from Eastern Europe in the decade ahead.

V

It is admittedly the case that the skeptic would have been a rather poor guide to the changes that have already occurred in recent years. But none of these changes have required the leap of imagination that a Soviet withdrawal from Europe must necessitate. In the absence of a quid pro quo by the West (one that is unlikely to be made, and that might well represent a bad bargain if it were to be made), a Soviet withdrawal would constitute an abnegation of power and interest that great powers do not freely undertake.

These brief considerations point to the scarcely startling conclusion that negotiating a general European settlement will prove, at best, a very difficult undertaking. It is not only that such a settlement must place in jeopardy Soviet interests heretofore, at least, regarded as critical; it will also require, if it is to have any real prospect of materializing, risk taking on the part of the West as well. The view that this need not be the case, or that if risks must be taken by the West they can still be kept modest, rests on the assumption that the Western states now hold almost all the high cards; this being so, they can force the Soviet Union to bear the burden of risk taking. But the advantage presumably conferred by Soviet domestic concerns can be overdone.

A European settlement of the kind discussed above would result in the substantial weakening, if not the virtual abandonment, of the Soviet military position in Central and Eastern Europe. That is, after all, what it is evidently intended to accomplish. Nor will it suffice to respond that legitimate Soviet security concerns will still be protected and that this is all Moscow is entitled to demand. For it is not only the case that Moscow's interests in Europe have not been limited to security in the narrower conventional sense, but it is also the case that even in this narrower sense the settlement proposed would have an adverse effect (though how adverse there is no way of knowing). Moreover, even if the Soviet Union were to keep the same force numbers in Europe, the new dispensation attending their presence would amount to the functional equivalent of their reduction. And in all likelihood the only feasible way of tightly garrisoning these forces would be to reduce their numbers drastically. If Moscow is nevertheless expected to accept this proposed settlement, its compensation for doing so will have to be a substantially reduced American military presence in Western Europe (and one, of course, that would be similarly garrisoned). The weakening of the Soviet position in Eastern Europe would thereby be compensated for by a weakening of the American position in Western Europe.

It is easy to see the attractions that an agreement of the above sort holds out. Not the least of these attractions is that in achieving such an agreement we would realize one of the long-standing objectives of postwar American foreign policy: the complete restoration of self-determination and independence to the nations of Eastern Europe. By formally agreeing to this restoration the Soviet Union would renounce the right, or rather the pretension of a right, to intervene in the internal affairs of these states. A line would thereby be drawn where there is today a persisting and dangerous uncertainty. Although this line might later be violated, the formal commitment to nonintervention would give an element of predictability that is now absent. A measure of stability would thereby be introduced into a situation that the momentum of change has rendered increasingly unstable. And while no one should pretend that the Soviet commitment to nonintervention would hold in any and all circumstances, it would impose a very substantial constraint on Moscow.

This is the attractive side of the new dispensation. There is another side, though, that must also be considered. The intent of the proposal is to substitute a sphere of influence for one of domination, to create, as it were, a new and more benign Yalta—indeed, the Yalta originally intended by the West. Would the new dispensation be temporary or permanent—a halfway station for the Soviets getting out altogether or for staying in and reconsolidating their position of power? The former seems by far the more plausible prospect. The obstacles to a reconsolidation of power are probably prohibitive. On the other hand, the halfway station is, for reasons earlier alluded to, almost inherently precarious and unstable. The Soviets must be expected to appreciate this. Their acceptance of the proposal, if not the outcome of sheer desperation, would be predicated on the quid pro quo already discussed, a substantial diminution of the American military presence in Western Europe.

The disadvantages of the new dispensation, then, are not so much the uncertainties attending its implementation in Eastern Europe. To be sure, these uncertainties—that the Eastern European states may insist on the departure of all Soviet forces; that the Soviets may, for whatever reason, break the agreement by intervening in a host state's internal affairs; and that, in any event, the agreement cannot be expected to freeze a situation that is inherently fluid and unstable—are by no means negligible. But they must be set against both the undesirable and the uncertain feature of the de facto situation that they are designed to redress. For the de facto situation is already one that is fluid and unstable.

The more serious disadvantages must be found elsewhere. An agreement of the kind here considered is almost certain to give rise to growing pressure for steps that will further weaken the American military presence in Europe. The mutual reduction of forces likely to be stipulated will initiate a process long called for by many in the United States. Even if the agreement did not provide for such a reduction, a highly unlikely prospect, pressure for reductions in the American military presence in Europe would mount because the agreement would ostensibly remove the circumstance that occasioned this presence in the first place. At least it would appear to do so as long as the Soviet Union faithfully carried out the terms of the agreement. Nor would this appearance lose its persuasiveness by virtue of the continued presence of Soviet forces in Eastern European states. Those forces, it would be said, now were closely circumscribed in function. And if Moscow gave solid indication of an intent to conform to the agreement, the argument would be persuasive. Considering the prospective benefits accruing from conformity, the Soviet Union might well decide to risk, if necessary, the interests that conformity could entail in order to capitalize on the benefits.

Nor would these benefits be limited to a weakened American troop presence in Europe. They would also include the heightened prospect of a denuclearized Central Europe. That prospect would merely be the logical consequence of a new dispensation in Eastern Europe that presumably removed the threat of Soviet armed intervention while providing as well for the reduction in the conventional forces of the superpowers. Indeed, these are the circumstances in which the arguments for denuclearization must appear particularly compelling to the Germans. In turn, the rising German pressure for denuclearization would provide added strength to those in this country pushing for further reductions in, or the withdrawal altogether of, the American force commitment in Europe.

If these considerations have merit, they point to very grave drawbacks to the proposal thus far considered for dealing with Eastern Europe. Are these drawbacks any greater than those attending the pursuit of alternative policies? It should be apparent that barring Western European political unity—an eventuality here ruled out for the decade ahead, but one the unexpected materialization of which would obviously render the speculation in these pages irrelevant—there are in principle only two broad alternatives to the approach considered above. One consists of some form of Austrian solution for Central and Eastern Europe, that is, the area's neutralization. The new status would be guaranteed by the Soviet Union and the Western Powers. The Eastern European states would maintain minimal military establishments and could entertain no political–military relationships with the West. They could have virtually any other relationship, however, and presumably would have close economic relations. Whether they would be permitted to join, or to have a special relationship with, the European Community may be left open.

What might lead the Soviet Union to give this alternative favorable consideration? Considered in isolation, an Austrian solution for Eastern Europe would seem to risk the sacrifice of virtually all Soviet interests and aspirations in Europe. Whatever residual control the Soviets might still exercise over Eastern Europe by maintaining garrisons there would now be sacrificed. With military power thus devalued in Eastern Europe, Germany's influence would soon become paramount for there would be little to counter its economic might.

These are the likely consequences of an Austrian solution, considered in isolation. But there is little reason to assume that the government of Mikhail Gorbachev, or any successor Soviet government, would be disposed to consider it in isolation. An end of the Warsaw Pact, the evident consequence of a neutralized Eastern Europe, would be considered only if attended as well by the end of the American military presence in Western Europe. This may not require the end of the Western alliance as such, but at a minimum it would surely require a return to the original meaning of the alliance—a unilateral guarantee by the United States of Western Europe. The formal alliance might persist, but the structure it has taken almost from the outset would be eviscerated.

The proposal of a neutralized Central and Eastern Europe is, in effect, a proposal to end the system of alliances that has dominated Europe for forty years and that, for all the criticism made of it, has given the old continent a remarkable era of peace and stability. If it is now contended that this system can in any event no longer hold out the same benefits, given developments in both Europe and the Soviet Union, the proponents of radical change must candidly confront the likely consequences of such change. It will not do to avoid facing up to these consequences by assuming that the rise of a new Europe will coincide with the decline of the old Europe, that a united Europe will fill the vacuum left by the Europe of alliance blocs. A united Europe that would fulfill the role and functions of the United States might eventuate in the course of the first two or three decades of the next century. To find it happening in the 1990s and to base American strategy, and the future of Soviet–American relations, on the expectation of its occurrence would be to run considerable risks.

Rather than a united Europe emerging to fill the vacuum created by the withdrawal of American power, the far greater likelihood would be a development that would imperil the prospect of Western European political unity. A Western Europe from which American forces had been withdrawn might well create the circumstances in which Germany would undo the ties of nearly half a century.

These sobering consequences can be dismissed as excessively pessimistic if it is assumed either that the Soviet Union no longer entertains the interests in Europe it has in the past or that even if it does continue to harbor much the same interests it no longer has the power effectively to pursue those interests. The first assumption must take the new thinking at face value, and more, while the second assumption must, if not simply write off military power, then at least find it increasingly subordinate in significance to economic power. Neither assumption can be taken for granted. If they should prove unfounded, the consequences of the schemes considered here could well lead to results that re-create dangers that forty years of effort were undertaken to avoid.

VI

If new departures incur the risks that they appear to incur, why not remain with the familiar? What is the great objection to the present course? That it underwrites uncertainty is clearly the case. On the other hand, it is far from apparent that the drawbacks of present policy can be avoided by any proposed alternative. Even if one assumes that the Soviet Union would agree to the first of the approaches considered above, the dispensation that resulted would not remove the uncertainty marking the prospects of Soviet military intervention in Eastern Europe. At best, it would lessen such an uncertainty. At the same time, it would open wider the door to prospects—particularly the recession of American power—the undesirability of which might well outweigh the benefits attendant on a greater measure of stability in Eastern Europe. Nor is it apparent that the growing dilemma of reconciling Germany's aspirations with the broader security interests of the Western Alliance would be any more susceptible of satisfactory resolution by such a new dispensation in Eastern Europe. Quite the contrary, the prospect of greater stability in Eastern Europe might, in the assurance it would give to Germany, only serve to sharpen this dilemma.

If the liabilities attending new departures in Europe are regularly minimized, it is largely because of optimistic assumptions respecting the extent of present Soviet weakness and vulnerability and the prospects for Europe's emerging political unity in the course of the decade ahead. Whereas Soviet need will presumably lead Moscow to accept Western proposals, though requiring a sacrifice of interests by Moscow, Europe's emerging political unity will provide a safeguard against liabilities attendant on these proposals. Neither set of assumptions, however, can be taken for granted. The more prudent assumptions are that Soviet need will not prove so great as to make Moscow a soft and pliable negotiating partner, and that Europe's political unity will remain an aspiration well beyond the 1990s. Given these assumptions, the attractiveness of new departures must diminish considerably.

To propose nothing better than the continuation of the present course is to incur the charge of Micawberism in policy. There is a difference, however, between passively submitting to events in the hope that something will turn up and eschewing alternative courses in the belief that they hold out greater risks then the course already being followed. The principal reason for avoiding the alternatives considered here is simply that the great desideratum of America's European policy must be to ensure that American power will remain engaged there for the foreseeable future. The most promising way of accomplishing this is the pursuit of the known.

3

The New Thinking and Its Limits
Soviet Foreign Policy under Gorbachev

Walter Laqueur

Important changes have taken place in Soviet foreign policy since 1985; this review addresses itself to the reasons for these changes, as well as to the question of their extent, and whether and to what degree they should be considered irreversible.

In the late Brezhnev era, as well as during the transition period after his death, the Soviet regime had maneuvered itself into a not-so-splendid isolation. True, strategic parity had been reached with America at an enormous price, but this (and conventional superiority) could not be translated into political gains. The war in Afghanistan went badly and was unpopular, and the country suffered other, albeit minor, setbacks in other parts of the world. The old foreign-policy doctrine became more and more obsolete; to enhance its security and to expand its sphere of influence the country had accepted a burden it could not indefinitely shoulder. Nor was it at all clear whether the heavy cost was in any way commensurate with the modest gains.

There was some heart searching even under Yuri Andropov, but no decisive action was, or could be, taken. Even during Mikhail Gorbachev's first two years the changes that took place were insignificant. The Twenty-seventh Party Congress (1986) gave Gorbachev a mandate for new foreign-policy initiatives, but foreign policy was not yet high on his list of priorities. True, there were indica-

Walter Laqueur • Center for Strategic and International Studies, Washington, D.C. 20006.

tions of a greater willingness to negotiate with Washington and to give up the isolationist orientation that had prevailed under Andrei Gromyko—who was replaced as foreign minister early in the new regime by Eduard Shevardnadze, a man more in line with Gorbachev's thinking. At the same time another stalwart, Boris Ponomarev (head of the Central Committee's foreign affairs department), was replaced by Anatoly Dobrynin, more conversant with the realities of world politics; but he too had been a pillar of Soviet foreign policy under Leonid Brezhnev. The speeches at the party congress and on subsequent occasions were replete with the traditional slogans about imperialism. If there was any new thinking, it manifested itself more in a new diplomatic style than in any radical new departures. Neither *glasnost* nor *perestroika* seemed to have deeply affected the general direction of Soviet foreign policy in 1986 or early 1987.

It was only toward the end of 1987, and particularly during 1988, that the new thinking became a more-or-less coherent system of thought and action. As we address ourselves to a description and an analysis of the new thinking, it ought to be noted that, as in other aspects of recent Soviet policy, the party line has been by no means uncontested. The new initiatives were not accepted by the Russia-firsters and the neo-Stalinists on the right who argued that they went too far, or by the radicals on the left who claimed that the changes did not go far enough. These disputes came out most clearly in the debates about past Soviet foreign policies, such as, for instance, the pact with the Nazis in 1939. The opinions voiced ranged from (almost) uncritical endorsement of Stalin's line as the only possible one in the given circumstances to (almost) total condemnation of the pact and its disastrous consequences. Similar divergences emerged with regard to responsibility for the outbreak of the Cold War, or, to point to a more recent example, the deployment of the SS-20 intermediate-range ballistic missile.

In a series of speeches in 1987 and 1988, Gorbachev and, above all, Shevardnadze, the new foreign minister, submitted Soviet foreign policy under Brezhnev to searching criticism. In a speech on the seventieth anniversary of the October Revolution, Gorbachev said that foreign-policy errors had been committed, opportunities that had opened up had not been made use of, and reaction to Western policy had been inadequate. Shevardnadze was more detailed and outspoken in his criticism. He said that during the preceding fifteen years the Soviet Union had lost its position as one of the leading industrial nations, and that as a result of this loss the country was unable to participate on equal terms in the political struggle. The Soviet Union had cooperated in material investments in impossible foreign-policy projects, resulting in enormous costs to the people.

But not everyone was willing to subscribe to such criticism. The differences concerned fundamental issues such as the demand for the demilitarization of Soviet foreign policy (strongly resisted by the military) or the priority to be given to the *all-human* over *class interests*. This new orientation was strongly opposed by the conservatives for both theoretical and practical reasons. The critics argued that it was not in conformity with traditional Marxist–Leninist doctrine; furthermore, it was bound to disorient Moscow's traditional allies abroad (the Communist parties in the West) without necessarily gaining new friends. The radicals, on the other hand, believed that such a reorientation was long overdue; foreign Communist parties when in power followed their own interests and could never be implicitly relied upon. Soviet hopes for achievements in the Third World had not been fulfilled on the whole; the influence of Western Communist parties and front organizations was steadily declining. The firmest friends, such as Cuba and Vietnam, were also the most costly. With their economy in shambles they had become accustomed to receive financial support on a more-or-less permanent basis.

But there was also criticism from the left for reasons of principle. When Shevardnadze claimed that Soviet foreign policy was deeply moral, some Moscow critics noted that a revolutionary revaluation of Soviet foreign-policy commitments and goals abroad had not yet taken place and that the Soviet Union still continued to support deeply unpopular regimes, not only among its immediate allies, but even among countries toward whom Moscow had no special treaty obligations. Thus, Soviet foreign policy had failed to take a stand on the brutal repression of the Chinese students' movement, and it had refrained from any comment on the Salman Rushdie affair. A prominent Moscow political commentator who had been upbraided by the Soviet foreign minister for expressing a critical view in this context vis-à-vis the Iranian government noted only half jokingly that soon Soviet commentators would not be permitted to make negative comments on any subject. Such an attitude on the part of the authorities could, of course, be defended on the ground of *realpolitik*, but it certainly did not bear out the claims of its "deeply moral character."

Given the fact that the new thinking, unlike the old party line, has not yet been canonized and spelled out in all details, its main characteristics can be defined only in broad outline.

THE NEW THINKING

Historical developments have made a basic redefinition of the Soviet national interest mandatory. The need for this redefinition is based on the assumption that in the foreseeable future neither the leading capitalist countries nor China is likely to collapse and that for years, perhaps for decades, the Soviet Union will be preoccupied with domestic reform to overcome its internal weaknesses.

It follows that Soviet strategic doctrine will have to be reexamined and changed. The arms race with the North Atlantic Treaty Organization (NATO) was a mistake because it could not result in decisive political advantages for the Soviet Union. The policy of threats and blustering vis-à-vis the West was counterproductive. It was also a mistake on the philosophical level because survival rather than victory should be the main aim of Soviet policy. Hence, the suggestions concerning the denuclearization of Europe and the reductions in force levels as suggested by Gorbachev on various occasions.

The long-term perspective of the new thinking is less clear. Some of the reformers argue, not without reason, that adopting the high moral ground gives the Soviet Union important advantages in the battle for Western public opinion. This refers to the championship of all-human values and interests, all-encompassing security, and opposition to nuclear deterrence as against the narrow class egoism of NATO, allegedly motivated by class hatred and scheduled to perpetuate the global social status quo. According to this doctrine the future still belongs to socialism cleansed from Stalinist deformations, even though capitalism has shown far greater dynamism and resilience than assumed by successive Marxist thinkers. Each country should have the right of freely choosing its future; chances are that most, if not all, will opt for some kind of socialist society.

Attaining these socialist societies, however, has now been moved into the very distant future. In the meantime the danger of nuclear war (let alone of conventional war) has not disappeared, it is argued, and even though priority should be given to a political approach in international politics, military power will remain a factor of paramount importance. The Soviet Union has to remain strong to deter the reactionary forces in the Western world. While the military capabilities of both sides should be lowered, the ratio of forces should remain as it is.

New thinking is criticized from the Soviet right for endangering the security of the Soviet Union and by the liberals because it does not go far enough. The reasoning of the right is, briefly, that new thinking rests too much on persuasion, on the attraction of the Soviet Union as a model for other countries. But given the difficulties facing the country, the attraction will be small for a long time to come. Historically, the Soviet Union (and Russia) has been more successful in building up a strong military machine rather than a strong economy. Furthermore, the Soviet Union (and Russia) has many outside enemies, and if national security is lowered they may be tempted to take advantage of what they perceive as Soviet weakness. Even if there should be no such hostility, the alleged threat of a cunning and powerful enemy cannot be discarded for domestic purposes. Deprivations and a dictatorial regime, as well as the traditional state-of-siege mentality, can be justified only if there is a clear and present outside danger.

Soviet liberals, on the other hand, argue that the reform in political thought does not go far enough. Soviet society is still deeply militarized (the indoctrination in schools and in the Communist youth organization); pacifism is a crime; the military is politically more influential than in any other developed country; conscription is longer than in all other European countries; and there is no meaningful civilian–parliamentary control on the budget, on the structure of the military establishment, or on weapons acquisition. At a time of domestic crisis the military could become a political factor of decisive importance, as it did in Poland for many years. Military leaders have not been in the forefront of the struggle for the new thinking or of perestroika. They have on occasion threatened, at least by implication, that severe cuts in their budget could have far-reaching political repercussions. They have had the support of the right wing in their demands and complaints. Serge Averintsev, one of the most respected Soviet intellectuals and a member of the Congress of Peoples' Deputies, has observed that the conservative party in the congress seemed to be frightened not so much by changes in Soviet domestic policy as by Gorbachev's foreign policy. Yegor Ligachev, for a long time the spokesman of the right in the party top leadership, had frequently emphasized the class approach in the conduct of foreign policy, pointing to the danger of losing traditional friends abroad. He could, until recent events, refer to East Germany, Bulgaria, and Czechoslovakia; the position of the embattled governments of these countries was not made any easier by Gorbachev's reforms; Vietnam was at most a half-hearted camp follower of the new thinking. Cuba and, until recently, Rumania, traditional mavericks, were dead set against the new policy. Among the major foreign Communist parties, the French and the Portuguese were opposed, whereas the Italians said that it did not go far enough.

The economic situation is a driving factor in the development of the new thinking. The budget deficit is now more than 100 billion rubles, shortages are growing in many fields, and the political consequences of these failures are dangerous in the extreme. In this situation additional revenue had to be raised and expenditure cut, except allocations for the poorest and most disadvantaged sections of the population. One of the obvious targets of these cuts was the huge military modernization and expansion program that had begun in the 1970s. For the first time an official figure was given with regard to current defense spending (77 billion rubles, 15.6 percent of the total state budget). This figure was more realistic than the previous official statistics, even though in the view of most Western (and some Soviet) experts it by no means covered all defense-related expenditure. Even earlier, in January 1988, Gorbachev had announced that by the end of 1991 the Soviet defense budget would be cut by 14 percent (and the production of arms and equipment by 19.5 percent). The demobilization of 500,000 officers and soldiers in 1989–1990 was also announced. The predictions made by Prime Minister Nikolai Ryzhkov went even further: military expenditure would be cut by almost one-half up to 1995. It might be premature to take these announcements at face value. But they certainly reflect the intention to relieve pressure on the Soviet Union's limited resources, to cut back inflation, and to reduce the huge budget deficit. It is not certain that these cuts, if and when carried out, will have the hoped-for effect in dramatically increasing investment in the civilian sector, boosting the output of light industry and food production, and thus relieving the shortage of consumer goods. But there can be no doubt that they are bound to have some effect. At the very least the Gorbachev reforms prevented a further growth in defense expenditure from the 16 percent he mentioned to 18–20 percent.

The economic situation and the urgent need for reforms led to a debate on rethinking strategic doctrine. The concept of reasonable strategic sufficiency was developed in 1985–1986 and became official policy (at least as a slogan) in the early summer of 1987. As far as the conduct of foreign policy is concerned, this meant the urgent need for relaxation of tensions, that is to say, a new détente. It meant cutting overdue losses, even if painful, with Afghanistan as an obvious example. It meant a dramatic improvement in atmospherics, much more far reaching than in the previous détente. In the short run it was less certain whether it meant any major Soviet concessions, political, territorial, or military. It was only natural that the Soviet leaders wanted to strike the best bargain, even from a position of relative weakness.

The changes in the style of Soviet diplomacy have been of considerable importance. In 1988 it was noted with regret that Soviet diplomats had become known in the past as *Messrs. Nyet*, who were on the whole ill informed and for whom compromise was a dirty word. It was realized that such a negative approach had resulted on occasion in short-term, local gains for the USSR, but it had not contributed to Soviet security and the international standing of the country in world affairs. To remedy this state of affairs the "charm offensive" was launched, most successfully by Gorbachev and his foreign minister, but also by officials of the foreign ministry and Soviet diplomats abroad. Not every member of the Politburo was as effective as Gorbachev, and the same is true with regard to the diplomats, especially those whose behavior and mental outlook had been formed in the Brezhnev era; but a brave effort was made by most.

At the same time as the changes in diplomatic style, relations between the foreign ministry and the KGB, which had been rather distant under Gromyko, became closer. Soviet diplomats were kept better informed and were given permission to be more outspoken in conversation with foreign interlocutors. Previously, they had often been at a disadvantage. The head of the Soviet delegation in the chemical warfare negotiations, Viktor Issraelyan, mentioned an extreme example of the *neglasnost* of the former period: for eight years his delegation had been forbidden to reveal whether or not the Soviet Union, in fact, possessed chemical and bacteriological weapons.

Soviet diplomats and foreign-affairs officials now appeared much more frequently in Western media and contributed articles to the Western press; at the same time Western diplomats were on occasion invited to appear on Soviet TV, radio, and the other mass media. Whereas in previous years the party line in domestic as in foreign affairs had been clear and unambiguous, under glasnost and the new thinking Soviet diplomats would air, on occasion, personal opinions or individual interpretations of the party line. This led to retractions and caused confusion among their foreign interlocutors. It is conceivable that sometimes such behavior was deliberate, but far more often it was the inevitable result of a more open political style. Thus, Western observers of the Soviet scene had to get accustomed to the fact that a highly placed Soviet official might advocate in a leading Soviet newspaper the establishment of a single world state and world government, but that his ideas would in no way commit the Soviet government.

To what extent has the Soviet foreign political agenda changed under Gorbachev and how successful has the Soviet Union been in attaining its goals? There have been certain shifts in priorities that have become basic changes. True, some of these priorities predate Gorbachev's rise. To give but three examples, Soviet attitudes toward terrorism (with a few notable exceptions) have become harsher, Soviet interest in the Third World has declined, and greater efforts have been made to reach a reconciliation with Beijing. These shifts were not sudden in coming; their origins lay in the early 1980s, even though they gathered momentum in the age of the new thinking. Some foreign observers believe that they have detected at various stages since 1985 a growing Soviet interest in Western Europe, and in Japan and the Pacific area. The Soviet leadership was said to have realized that new centers of power in world affairs had emerged in the last decade or two and that the Soviet–U.S. relationship was of lesser importance, therefore, than in the past. This was true inasmuch as the interests of a superpower are, a priori, unlimited, in that it cannot afford to ignore any part of the globe, however remote.

It is also true that Gorbachev and his supporters have paid particular attention to and invested special efforts in activating Soviet policy in Western Europe. This is so for some obvious reasons: Europe's steadily growing economic importance and the prospects of even greater unity after 1992. But there was also the fact that sympathies for perestroika were very pronounced in Western Europe; Gorbachev was more popular than President Ronald Reagan in most West European countries. His popularity in some countries even exceeded that of the local leaders. This popularity and the attractiveness of the slogan about the common European home led some outside observers (wrongly) into discovering a basic reorientation of Soviet foreign policy, away from the preoccupation with the United States and toward a rapprochement with Western Europe. The geostrategic position of the Soviet Union was stressed in Moscow on various occasions as a bridge that would bring the European and Far Eastern industrial societies closer to the Soviet markets. But Japan and the other highly developed Far Eastern economies had never been particularly interested in the Soviet markets, and after some initial professions of interest, the Western European response proved disappointing. It will be recalled that the overall trend of trade between the market economies and the Soviet Union has been declining since the late 1970s, and the loans and credits extended by foreign countries could achieve little more than halt this decline and, in the case of some countries, bring the level of trade back to the level of the mid 1970s.

The reason for the decline was not ill will on the part of the West and Japan, but the Soviet inability to pay for Western (and Far Eastern) imports in hard currency or to export commodities in demand in the West. The Soviet position was further harmed by the decline in world prices of oil and gas, the most important items in the Soviet export list. The world price of oil began to fall just nine months after Gorbachev's appointment, and it is estimated that the Soviet Union lost as a result of the fall not less than $60 billion between early 1986 and the end of 1988.

The idea of common ventures had been discussed for a long time, but with a few exceptions the results were meager. Despite repeated promises to simplify procedures for foreign firms operating in the Soviet Union, there was very slow progress in this direction. Many Western experts reached the conclusion that the conditions for major common enterprise did not exist and that greater service could be done to the Soviet economy by providing training to Soviet managers, as Chancellor Helmut Kohl, among others, had promised. On occasion, the idea of a Western Marshall plan was voiced; the term was bound to be insulting to Soviet ears, nor was it likely that the West and Japan would be able and willing to find the enormous financial resources commensurate with Soviet needs. A widespread view in the West was that even investment in the magnitude of tens of billions of dollars (or rubles) would not be effective; the Soviet Union, after all, had invested hundred of billions in agriculture without achieving significant results. In brief, the prevailing view was that short of systemic change in the Soviet economy there was very little the market economies could do to help. Individual enterprises might engage in mutually beneficial projects, but the overall picture was unlikely to change in the foreseeable future. This view was challenged by West German government opinion. But this view may owe more to political interest than to a different economic assessment.

It ought to be mentioned in passing that concurrently a bitter dispute took place inside the Soviet Union about the desirability of making use of Western credits. The ideologists of the Russian right have claimed that borrowing from the West would have the most deleterious consequences, suggesting that the International Monetary Fund, the World Bank, and other such international (and in their eyes sinister) organizations would eventually dictate Soviet politics, and that the Soviet Union would turn into something little better than a banana republic.

Unorthodox economists like Nikolai Shmelyov have argued, on the other hand, that the credit standing of the Soviet Union is high, and that unless the Kremlin leadership imported significant quantities of consumer goods, the domestic situation would further worsen, in the absence of incentives to work harder. Short of a major infusion of imports of this kind, the vicious circle besetting the Soviet economy would never be broken and the whole reform movement would fail. Most Soviet policymakers have taken positions somewhere between these two extremes. They have been in favor of taking some loans and importing some consumer goods but have opposed a buying spree and have been wary of growing dependence.

Although the Soviet Union has tried hard to encourage Western economic involvement in the Soviet Union and Eastern Europe, its attitude toward the EEC (European Economic Community) and, in particular, the idea of a common European market has been less than enthusiastic. The Soviets realized that the common market was an established fact and that it was likely to strengthen Europe's role in the world. The old Soviet position of relentless opposition had, therefore, to be modified. But the lack of enthusiasm was still understandable because the position of CMEA (Comecon) in the Eastern bloc will become even weaker, export of Soviet manufactured goods even more difficult, and the possibility of playing one major European country against another more limited. True, there is the chance of growing economic conflict between the EEC and the United States, but this would not be of immediate benefit to the Soviet Union.

Among Soviet achievements the INF (intermediate-range nuclear forces) agreement should be mentioned because it greatly enhanced Soviet popularity in Western Europe. In addition, relations with Social Democratic and Green parties in Western Europe have become much friendlier. This followed from the realization in Moscow that all Western European Communist parties were in a state of decline, some more than others, and that doctrinally the Gorbachev reforms had as much in common with the Social Democrats as with the pro-Stalinist Communist parties. Relations with the Greens were more complicated. On the one hand, among the Greens there was and is a considerable reservoir of good will toward the Eastern bloc, and also a common interest: to keep Western European defense budgets down. On the other hand, the Soviet Union and the other East European regimes have been among the world's most serious offenders against ecological concerns, and their belated willingness to make amends was not wholehearted. The Greens could not possibly ignore this.

This change in mood should not be underrated. But it was not what the Soviet Union needed most. Hence, the particular attention devoted to the country in which sympathies for Gorbachev were greatest, and which also happened to be the strongest economically—West Germany. Gorbachev's visit to Bonn in June 1989 was a personal triumph of a high order, and the speeches about a new age that had opened in the relations between the two countries were no doubt sincerely meant. It was indeed a far cry from the Soviet suspicions of German revanchism, voiced even a few years earlier, to Bonn's new status as privileged interlocutor.

There were, however, political problems confronting the two countries that could not be overcome at the present stage by ardent professions of good will and sympathy. First and foremost was the East German issue: the concept of the common European home was bound to remain an abstraction as long as there was no progress toward German reunification. The East German regime vehemently opposed any such move. The country that had been believed for a long time to be the strongest inside the Soviet bloc suddenly showed signs of great weakness.

As the year 1989 approached its end, the East German Communist regime disintegrated, and at the same time non-Communist forces came to power in Hungary and Czechoslovakia. Even earlier a new government under the leadership of Solidarity had taken over in Warsaw. The Soviet empire in Eastern Europe had virtually ceased to exist. According to the Brezhnev doctrine, first formulated in a speech in Warsaw in November 1968, the Soviet Union was committed to intervene, militarily if necessary, whenever the so-called gains of socialism were threatened in Eastern Europe; in other words, Communist rule was irreversible. The Soviet leadership could have intervened in 1989, and it probably would have succeeded in propping up the various Communist regimes. But it would have meant no more than postponing the day of reckoning by a few years. Communism in Eastern Europe had failed; if forty-five years after the end of the war they were not in a position to survive without Soviet military help, the game, from a Soviet point of view, was no longer worth the candle. Economically, no advantages had accrued to Moscow from its presence in Eastern Europe since the middle seventies; from a strategic view Eastern Europe was no longer thought essential.

The Soviet retreat from Eastern Europe has been the most important development in Soviet foreign policy so far, and it happened not so much because of the new thinking but because of the momentum of events in this part of Europe. It does not mean that the Soviet Union no longer has security interests in Eastern Europe. It will insist on arrangements that will safeguard these interests.

CHINA AND THE SOVIET UNION

The Sino-Soviet conflict was rooted in territorial disputes about spheres of influence, but at its core were the issues of independence and ideological authority concerning the structure of a Communist society. While the Soviet Union stagnated, China carried out economic reforms in the early 1980s. This did not prevent a reduction of tensions; in 1982 Brezhnev suggested reconciliation with China, which, he declared, was after all a socialist state. While there were few tangible advances toward reconciliation in the political field, economic relations again expanded; in 1986 trade was ten times what it had been in the 1970s, and Soviet experts, albeit in small numbers, returned to China. When Gorbachev came to power he suggested that China should restore interparty relations, having declared his willingness to make far-reaching concessions with regard to China's three main complaints (Soviet involvement in Afghanistan, the Vietnamese occupation of Cambodia, and the Soviet military buildup along the Sino-Soviet border).

Thus, a normalization of relations slowly began, reaching its culmination with Gorbachev's visit to China in May 1989, the first Sino-Soviet summit in thirty years. This meeting was not, however, an outstanding success; it coincided with major demonstrations in Beijing and in other cities that resulted in bloody repression. While previously the Chinese reforms had evoked interest and occasionally even admiration in Moscow, it now appeared that they had led to inflation and widespread corruption. China, in the words of one observer, had not succeeded in establishing market socialism but had combined some of the worst features of capitalism and socialism in a dismantled planned economy with uncoordinated, unlinked, and imperfect markets for individual goods. Political and social reforms had lagged behind economic reform. The Chinese leadership was overage and tended to excessive caution. In view of the deep internal crises facing the Chinese leadership, the Soviets must have reached the conclusion that China was much less of a model, and also much less of a threat, than had been assumed in earlier years, and this, paradoxically, made further normalization possible. Gorbachev decided to reduce Soviet Far Eastern forces by 200,000 troops.

The withdrawal of Soviet troops from Afghanistan and the Vietnamese retreat from Cambodia were completed in 1989, and thus the preconditions were created for a further rapprochement between the two Communist superpowers. However, it still remained improbable that Soviet concessions would lead in the foreseeable future to truly close cooperation. The Communist regime in China was almost totally preoccupied with internal affairs—the struggle for a new leadership and a broadly agreed-upon domestic policy to cope with the severe crises facing it. In these circumstances foreign policy was bound to take a secondary place. And the Soviets realized that it was pointless to bring pressure on the Chinese leadership to force the pace of reconciliation, for they could not be certain whether their interlocutors in Beijing would still be in power a few years hence.

In any case there were limits to Chinese enthusiasm for a new détente with Moscow; the recollections of the old status of inferiority and dependence still rankled. Nor was the Soviet Union, given its economic weakness, a very attractive trading partner. In the words of a highly placed Chinese official: "They send us low-quality wood and we send them second-class canned food." True, the fact that the Soviet Union, unlike the Western powers, did not utter any criticism of Chinese repression after the events of Tiananmen Square did not fail to make a favorable impression in Beijing. But this was hardly sufficient to restore the old spirit of Communist solidarity. The common Marxist–Leninist–Stalinist heritage no longer provided sufficient cement for a firm alliance between the two powers.

The most the Soviet Union could hope for during the interregnum in China (and probably well beyond) was a further reduction of tensions.

THE THIRD WORLD AND THE SOVIET UNION

Soviet policy in the Third World had been very active during the 1960s and 1970s. Friendship treaties were signed, considerable quantities of arms supplied, and great efforts invested intensifying economic and cultural ties. On the ideological level, theorists claimed that the camp of countries of socialist orientation was steadily growing. But even before Gorbachev, there were clear signs of dissatisfaction in Moscow with the results of Soviet Third World policies. Only under glasnost, however, were these policies submitted to searching analysis.

The result was largely negative. It was admitted in retrospect that although the idea that Third World countries would opt for a noncapitalist road might be correct on the theoretical level, it was irrelevant under present conditions. The main problem, as one Soviet observer put it, was that in most of these countries there was not enough capitalist development, rather than too much of it; they had stagnated, and over the years in some countries output had even declined. Their socialism had been largely a matter of revolutionary phraseology to cover all kinds of religious and nationalist doctrines, tribal ambitions, and xenophobia. The working class in these countries was not yet politically literate, and the military and the intelligentsia that usually constituted the leading stratum followed their own group interests or individual ambitions.

In brief, the Soviet investment had not brought worthwhile returns. If Moscow had intervened in a regional conflict, it had antagonized one side without being able to count on the lasting gratitude of the other. The arms supplied had not necessarily led to a growth of Soviet influence, and more often than not they had not even been paid for in the end. Economic links had sometimes been profitable, but more often they had resulted in losses for the Soviet Union. The training of Asian and African students and technicians in the Soviet Union had often caused friction and disappointment. Furthermore, Soviet investment in Third World countries had been unpopular at home. It was asked with growing frequency if Russia could really afford to extend help to underdeveloped countries if the situation on the domestic front was so unsatisfactory and the question whether good use would be made of Soviet assistance so uncertain.

All this heart searching did not mean that Soviet policy in the Third World would come to a standstill. The foreign minister visited New Delhi as well as some Middle Eastern capitals. There was talk of a possible role for the Soviet Union as an arbiter in the South African conflict. The Soviet Union wooed Iran as well as Iraq and insisted on participating in any Arab–Israeli peace settlement. Soviet diplomats continued to be active in South Asia, and the Soviet presence in the Pacific region was strengthened. It ought to be mentioned in passing that there was little progress in relations with Japan, except an agreement that Gorbachev should visit Tokyo in 1991. From time to time the possibility of Soviet territorial concessions was mentioned, but these were mere rumors for the time being. It was apparently realized in Moscow that a fundamental reorientation of Japanese policy toward the Soviet Union resulting in massive new investment was unlikely. In these circumstances concessions seemed not to be called for.

Soviet diplomacy was more active in Latin America than in previous years, and the Soviet Union's commitment to Cuba and Nicaragua did not decrease. Shevardnadze visited various Latin American capitals, and Brazilian, Argentinian, and other Latin American leaders came to Moscow. While there were no more illusions in Moscow with regard to the revolutionary potential of Latin America, and while sympathies for Cuba and the former Sandinista government in Nicaragua had shrunk all over the continent, it was also clear that traditional suspicions and enmity toward the United States persisted. In the circumstances a strengthening of the Soviet position in Latin America seemed to be politically advisable, and in any case not very costly. The intention was not to challenge frontally the American positions or to try to deny the United States access to strategically important areas and minerals; this would have merely complicated Soviet–American relations without giving any commensurate benefit to the Soviet Union. The Soviet presence in Latin America was a long-term investment in an area of high U.S. sensitivity and vulnerability. There was no intention to provoke the United States, but the very fact of the Soviet presence in the hemisphere was to demonstrate that the Soviets could not be ignored. Soviet strategy in Latin America was low key and cautious, encouraging a drift away from U.S. influence, but only in a discreet way. On the other hand, Soviet ability to offer a true alternative to the United States (and the EEC) was exceedingly limited in Latin America. To a limited extent the Soviets could provide a political and strategic counterbalance, but no market for Latin American products.

THE UNITED STATES AND THE SOVIET UNION

No mention has been made so far of Soviet strategies toward the country still considered the most important, the United States. Compared with the relentless hostility and the dire threats of the early 1980s, the Soviet shift was most striking. After an uncertain start (the Geneva and Reykjavik summits) the Reagan–Gorbachev meetings in Washington and Moscow, as well as the various meetings between the foreign ministers, resulted in agreements on a wide variety of topics: confidence-building measures, intermediate-range nuclear arms, nuclear testing, and chemical weapons. Progress on agreements involving strategic nuclear and conventional arms was also forthcoming.

Understandings were reached regarding certain regional issues, and there was cooperation on issues of common concern to both superpowers (nuclear proliferation, terrorism, and ecological problems). However, Soviet foreign policy was still torn between the desire not to be drawn by one of their less predictable clients, such as Libya or Syria, into foreign political adventures and the wish not to give up positions acquired in these and other countries. The Soviet Union wanted to be at one and the same time a responsible, dependable, peace-loving power and the friend of not-so-responsible Third World leaders such as Muammar Khadhafi.

Nor were long-term aims, such as the weakening of NATO, affected by changes in Soviet foreign policy. Thus, the INF treaty was explained for domestic consumption as weakening the military presence of the United States in Europe, reducing the value of U.S. nuclear guarantees so that eventually Western European leaders would have to ask themselves what value NATO still had in the circumstances. There was no lessening of the Soviet intelligence effort, nor was the idea of the ideological struggle given up. What was new was the realization that political goals ought to be pursued primarily by political means, that the political offensive had to appeal beyond opposition groups in the West, and that, generally speaking, Soviet foreign policy should not expect a dramatic change in its favor with regard to the global correlation of forces so often invoked in previous years.

CONCLUSIONS

Most of these changes are connected with Gorbachev's perestroika policy on the domestic scene. What if the reform movement should stall or go into reverse? The links among reform, glasnost, and foreign-policy new thinking are obvious, and it would be premature to regard the new orientation in Soviet foreign policy as irreversible. Sections of the leadership of the Soviet Communist party and also of its middle and lower echelons feel somewhat uncomfortable with rapprochement with the West, which, they claim, has gone too far. They suspect Western hostile intentions and believe that even if America and Western Europe do not harbor such designs, Western civilization (mass culture) has a negative influence on Soviet youth. Foreigners, they believe, have always hated Russia and wished to harm her; Soviet policy should therefore keep its distance, as it has in the past. These same circles, however, by no means favor an alternative orientation, towards the Third World, China, or the Warsaw Pact states. If their influence in the Politburo should prevail, this will manifest itself in something akin to a neoisolationist strategy. It is unlikely that they will adopt an aggressive foreign policy, other than making use of opportunities that obtrude themselves; they are aware of the fact that extending the Soviet sphere of influence would not bring about a lessening of internal tensions, such as the ferment among Soviet nationalities; on the contrary, it might aggravate the state of affairs. Radical reformers as well as conservatives and neo-Stalinists agree on at least one point—that the domestic situation is more threatening and that it ought to be given priority on the national agenda.

What long-term changes can be realistically expected in Soviet foreign-policy orientation? The idea of world revolution was given up long ago, but the Soviet Union has continued to consider itself until recently the leader of the socialist camp (or, to be precise, part of it). A further lessening of ideological motivation in Soviet foreign policy seems certain over the years to come, be it only because the continued existence of the Soviet Union within the old borders and on the basis of the old relationships has become uncertain. From a long-term perspective, Soviet ideology will continue to develop, perhaps towards a populism that can be defined with equal justification as either left or right. A further deideologization is also likely, but it should be borne in mind that traditionally Russia, and later the Soviet Union, has experienced a greater need for such articles of faith than have the United States, Western Europe, and, indeed, most other nations. It is not certain that this need will change in the foreseeable future.

4

Arms Control and the Future of Nuclear Weapons

Steven A. Maaranen

Arms control, to include both actual treaties and the arms-control process, has had a notable impact on U.S. nuclear weapons over the past twenty-five years or more. The influence of arms control is likely to continue and, in combination with other factors, to contribute to a substantial reorientation of future U.S. military strategy with narrower, though still crucial, roles for nuclear weapons. In particular:

Already, the United States and the Soviet Union have agreed to eliminate intermediate-range nuclear forces, have agreed to discuss reductions in short-range nuclear weapons, and are nearing completion of negotiated reductions to more equal, comparable, and stable strategic nuclear force levels. Such forces possessed by the superpowers will be highly deterring, but less credibly applicable to actual military plans and political uses.

Subsequently, U.S.–Soviet arms control seems destined to support a major and deliberate movement to relegate nuclear forces to primarily a retaliatory deterrent role, accompanied by a shift to forces designed to be stable above all, even at the sacrifice of attributes useful for current types of military missions. This shift will occur in part by design (it has been an objective of U.S. arms-control policy for decades), in part by the force of technological change (the increasing deployment of highly survivable nuclear forces, and the growing capabilities of nonnuclear weapons and defensive systems), and in part by changing global circumstances (the shift from a bipolar to a multipolar world).

Steven A. Maaranen • Center for National Security Studies, Los Alamos National Laboratory, Los Alamos, New Mexico 87545.

If future negotiations contemplate reductions far beyond those of the Strategic Arms Reduction Talks (START)—say, to a few thousand nuclear weapons—U.S. and Soviet forces would be drawn down to the level where the British and French nuclear forces, and the forces which the People's Republic of China (PRC) might deploy by that time, would be too large to be ignored. This situation would force serious consideration of the inclusion in negotiations of the nuclear forces of all important nuclear powers (at least France, the United Kingdom, and the PRC).

Finally, interest is likely to continue for many years in approaches to nuclear arms control that differ from direct, quantitative limits or reductions. Several approaches have been pursued: limits on the development of new technology, such as nuclear-testing constraints; nuclear-weapon-free zones; and naval nuclear bans. Some peripheral agreements of this kind, for example, the South Pacific Nuclear Free Zone, could be concluded. But because such approaches would have such widespread and unpredictable or unmanageable consequences for the national security planning of the major nuclear powers, they are unlikely to replace the more orderly and familiar Strategic Arms Limitation Talks (SALT)–START–Intermediate Nuclear Forces (INF) approach.

THE ARMS-CONTROL CONTEXT

For the past forty years, U.S. defense strategy and national security policy have relied on the powerful deterrent effects of nuclear weapons to prevent large-scale nuclear war, and to prevent or dampen lesser conflicts, especially those that might occur between nuclear-weapon states. In implementing this policy, the United States has constructed the concepts of deliberate escalation and nuclear first use, within the strategy of flexible response, to extend nuclear deterrence from the U.S. homeland to the territories of its friends and allies. At the same time, the United States has endorsed the concept of strategic stability and has striven to sustain a stable nuclear balance with the Soviet Union so that deterrence is ensured and no prospect can arise for one side to employ nuclear weapons successfully against the other nuclear superpower. In practice, the goal of stability has been at odds with the underlying requirements of flexible response, since flexible response requires the ability to credibly threaten the first use of nuclear weapons, and to engage in, and control, nuclear escalation. This dual approach to national security policy has meant retaining a powerful and diversified U.S. arsenal to underwrite all elements of U.S. strategy, while seeking arms-control negotiations and agreements that will provide both military and political stability at reduced levels of nuclear arms.

In principle, the objective of stability has been predominant in U.S. arms-control thinking, but negotiating outcomes have not moved very far in that direction. SALT I and II represented an effort to freeze the arms race by ensuring the dominance of offensive over defensive forces, accommodating to Soviet nuclear parity, and forging a new political relationship based on détente; all the while protecting critical elements of the U.S. nuclear force and future force plans. That approach was a failure through the early 1980s in that the Soviets failed to develop and size their forces and operate by doctrines in accordance with the American concept of stability. Instead, the Soviets proved more willing than the United States to invest in and compete with new nuclear forces, and Moscow sought to use the coercive power of nuclear weapons as a means of achieving Soviet geopolitical ends.

The U.S. commitment to seeking strategic and geopolitical stability through arms control did not fade, however. The Reagan administration was determined to meet the relentless Soviet political–military challenge that proved to be peaking in the late 1970s and early 1980s by closing the window of vulnerability through strategic force modernization. However, it deliberately refused to commit itself to reestablishing U.S. nuclear superiority. And now, in view of the Soviet military recession, the Bush administration is evincing an even greater interest in achieving a stable strategic policy and U.S.–Soviet relationship. Indeed, stability has arguably now become the chief objective of both U.S. arms-control policy and nuclear-force procurement programs.

A great deal is changing in the context and content of nuclear arms control; however, the central themes of U.S. policy, as well as the key conditions that affect that policy, are proving quite durable. First of all, the Soviet Union, even as it shifts somewhat from the center stage of U.S. political and military concerns, will remain in raw military terms the world's only other nuclear superpower, and hence it will be a fundamental U.S. security policy concern, as long as the USSR remains a viable political entity.

Next, the accustomed practice of the United States to use the arms-control process for addressing key issues in U.S. national security is not changing. Many groups with an interest in shaping national security policy find the arms-control process to be an attractive forum, and much of the U.S. public remains wedded to the idea that arms control may provide an avenue for creating national and international security, either in combination with or as an alternative to military policy and diplomacy.

Similarly, the dramatic changes of the past few years have not altered the goals of U.S. nuclear arms-control policy. The United States has sought through arms control to achieve rough numerical parity at lower levels of nuclear forces; strategic stability, particularly by limiting the number of vulnerable, prompt counterforce weapons on each side; qualitative restraints, but only on destabilizing technologies; effective verification of compliance; and Soviet acceptance that these are the principles that should govern the force structures, policies, and strategic relationship for both sides. Boundaries on movement in this direction have been set by U.S. and Soviet operational military requirements, and by Soviet reluctance to accept the validity of the U.S. approach to military strategy.

Finally, the United States continues to seek to integrate its strategic offense, strategic defense, and space arms-control policies, at least loosely, in pursuit of its goals. In the 1970s the United States sought to establish deterrence based on the concept of mutual vulnerability and negotiated the SALT I package of interim offensive restraints and the Anti-Ballistic Missile (ABM) Treaty in support of that concept. Since the early 1980s the United States has sought to create an option to achieve stability and deterrence in a different way in the future. Instead of ensuring mutual vulnerability by attempting to negotiate smaller, balanced offensive arsenals and to ban defenses, the United States has explored the possibility of supplanting or supplementing offenses with an increasing contribution of strategic defenses over time. The three-part nuclear arms-control negotiation— START, INF, Defense and Space (D&S)—was developed in the 1980s to advance this new agenda.

STRATEGIC NUCLEAR ARMS NEGOTIATIONS

In START the United States has sought stability and verifiability at significantly reduced numbers of strategic nuclear forces. The U.S. government has argued that these reductions would contribute to U.S. security even in the absence of strategic defenses or a Defense and Space treaty and could usefully be merged with defenses to provide the transition to a new national strategy that would provide a more secure and stable form of deterrence in the future.

The START treaty, which is nearing completion, would institute a process of strategic nuclear force reductions that would proceed for a period of seven years. Thus completing the negotiations and implementing the treaty could impose a framework on the sizing and modernization of U.S. forces that could last for a decade or more. At the U.S.–Soviet summit in July 1990, however, the sides agreed to enter into a new round of negotiations as soon as START is ratified. If those negotiations proceed rapidly, it is possible that U.S. and Soviet force drawdowns will proceed even faster, and further, than will be provided for in START.

The U.S. government hopes through the conclusion of an equitable and effective START treaty to improve political and military relations between the United States and the Soviet Union. The conclusion of an agreement should allow for stabilization of the arms race by capping the numbers and types of weapons on both sides and reducing incentives or pressures to build larger forces. This should place strategic deterrence on a surer, more stable footing and provide an effective, agreed-upon basis for future negotiated reductions. The treaty will limit each side to no more than 1,600 deployed intercontinental ballistic missiles (ICBMs), submarine-launched ballistic missiles (SLBMs), and heavy bombers, while the number of accountable warheads on these delivery systems will be limited to no more than 6,000.

In terms of the finer structure of a START agreement, the United States is seeking, through agreement on sublimits governing various types of nuclear delivery vehicles and the warheads they carry, to enhance crisis stability; that is, the United States wants to reshape the Soviet strategic arsenal to make it less of a first-strike threat and to reduce its attractiveness as a first-strike target.

First and foremost, the United States wants to place a warhead sublimit of 3,000 to 3,300 on deployed ICBMs, which are argued to be particularly dangerous and destabilizing. Of the ICBMs, the 308 Soviet fixed, heavy, land-based missiles carrying ten multiple, independently targetable warheads (the SS-18s) are a particular concern. Each is accurate enough to destroy several U.S. ICBMs and flies quickly to its targets, but these counterforce missiles are themselves vulnerable to attack. The Soviets have agreed in the negotiations to reduce these heavy ICBMs by one-half, to 154 missiles carrying no more than 1,540 warheads. For the remaining ICBMs, the United States favors some shift from fixed to mobile basing to improve their survivability, and the deployment of up to 1,100 mobile ICBM warheads has been agreed upon.[1]

Because of their inherent survivability, SLBMs are viewed by the United States as more benign or crisis stable than ICBMs. The United States proposes a sublimit of 4,900 warheads on the combination of deployed SLBMs and ICBMs. While the ICBM sublimit would force the Soviets to a smaller fraction of ICBMs in their force, the 4,900 sublimit would allow for an even larger shift from land-based to sea-based missiles if either country so desired. That is, a side could choose to deploy all of its 4,900 ballistic missile warheads on SLBMs, but no more than 3,000 to 3,300 of its 4,900 warheads on ICBMs.

Finally, the U.S. proposal is designed to encourage bomber and cruise missile forces, on the basis that they are slow flying and cannot be used for a short-warning attack. Thus, they are argued to be good weapons for retaliation but unsuitable for a first strike and hence are crisis stable. The 4,900 sublimit on ballistic-missile warheads would encourage the sides to deploy at least 1,100 of their 6,000 accountable warheads on bombers. Deployment of bombers is also encouraged by the provision that heavy bombers carrying only gravity bombs and short-range missiles count as one delivery system under the 1,600 limit, and as only one warhead, even though they can in reality carry many more bombs. Moreover, heavy bombers equipped to carry long-range cruise missiles are treated in such a way that the number of warheads accounted against them may be less than the number they actually carry.

Finally, sea-launched cruise missiles (SLCMs) will not be limited at all under the terms of the treaty, although each side will annually declare the number of SLCMs it plans to deploy each year, and this number is not to exceed 880.

All in all, these sublimits, applied to the forces of both sides, would bound the asymmetry between the U.S. and Soviet force structures, reducing the potential first-strike capacity of the Soviet arsenal and providing for a more survivable set of forces on both sides.

While the U.S. Joint Chiefs of Staff have agreed that they can provide an effective U.S. national strategy under the U.S. START limits, assuming adequate modernization of the forces, they have not indicated exactly how they will choose to structure the nuclear forces, that is, how many nuclear-powered ballistic missile submarines (SSBNs), how many of what types of ICBMs, and how the bomber force would be configured. However, because of decisions and investments already made, there are bounds on likely outcomes for U.S. strategic force structure.

- For the bomber force, the START treaty will result in favorable counting rules for both penetrating bombers and air-launched-cruise-missile-carrying bombers, although the rules will not be as favorable as the United States initially sought. Hence, there will be an incentive to deploy as many penetrating bombers as possible, and also to deploy a sizeable cruise missile force. Expected Congressional cuts in the B-2 *Stealth* bomber force procurement will reduce the number of U.S. penetrating bombers, as well as the actual number of weapons we will be able to deploy under START. On the other hand, a slowdown in modernization of Soviet air defenses, possible under their changing defense program, could allow the B-1B bomber force to be effective as penetrators longer than is now expected.
- For the submarine force, the *Poseidon* C-3 and C-4 boats will phase out of the inventory in the mid- to late-1990s, as they are replaced by *Trident* submarines. That force will probably be composed of at least the eighteen *Tridents* that have received funding so far. If unmodified (such as with fewer missiles per boat or fewer warheads per missile), or if the boats in long-term refit are not excluded from START weapon counts, these eighteen submarines would carry 3,456 accountable weapons and 432 launchers under START rules.
- The structure and size of the ICBM force is less certain. The United States currently sustains a sizable silo-based force because of its high alert rate, low operating costs, flexibility, and responsiveness, especially for limited strategic attack options. The need to modernize the silo-based *Minuteman* II and III force, and the U.S. desire to place at least a part of its land-based warheads on survivable missiles, has led the United States to continue to develop the option to deploy some number of mobile single-warhead, land-mobile *Midgetmen* and/or ten-warhead, rail-mobile *Peacekeepers* in the 1990s.

In fashioning a START force structure, the United States will have several choices within and between the elements of the strategic triad. Whatever changes are finally made, the resulting U.S. force is almost certain to include a larger fraction of survivable and/or slow weapons than the current force (counted under START rules), signifying a shift in U.S. nuclear forces toward more survivable, stable weapon types and force size and mix. However, with reduced budgets and greater reluctance to build the next generation of strategic nuclear systems (B-2, rail-mobile *Peacekeeper*, *Midgetman*) the speed of the shift toward survivable systems in the United States could be slowed.

The START treaty proposal will also result in modest progress toward the U.S. arms-control goal of reduced numbers of nuclear weapons. There would be approximately 50 percent reductions in treaty-accountable forces to 6,000 nuclear weapons and 1,600 delivery vehicles. Counting rules and nonaccountable weapons would significantly raise the total of actual weapons to perhaps 8,000 to 10,000. Still, the START reductions and sublimits would represent the first negotiated cuts in strategic nuclear arms and would require large reductions from current U.S. and Soviet weapon levels. In fact, many analyses suggest that the START proposals would bring U.S. (and probably Soviet) nuclear forces down to the level where they could just meet their current respective targeting requirements. That is, the resulting size and composition of U.S. forces would be such that the United States probably could avoid fundamental changes in its nuclear strategy and could continue, as today, to carry out the nuclear elements of the flexible-response strategy. However, because these reductions could nearly eliminate the margin of error or residual forces that may improve confidence that plans can be carried out successfully, the START agreement could make the calculation of success in an attack even more difficult than it is today. This could strengthen deterrence and stability but also make nuclear forces marginally less applicable to actual military plans and political uses than today.

Still, at the June 1990 summit, the United States and the USSR did commit themselves to the negotiation of "further stabilizing reductions." A number of unofficial ideas have already emerged in the United States for strategic arms reductions subsequent to a START agreement. Prominent among these is a further 50 percent cut. Also, Soviet President Mikhail Gorbachev has proposed a further 50 percent reduction followed by the elimination of all nuclear weapons by the year 2000, and groups of Soviet scientists and institute scholars have proposed reductions to 500–600 single-warhead mobile missiles on each side. Reductions of this order, while unlikely to receive serious consideration in the near term, may influence the longer term arms-control agenda and create an expectation of continuing strategic nuclear force reductions.

The United States has made it clear that it will place survivability and stability at the top of its list of objectives for the next round of START. But significant numerical reductions are also likely, and these could lead the United States to important changes in both national strategy and nuclear policy. How deep could such reductions be? The number of nuclear weapons required to implement a given nuclear policy is driven primarily by the number and type of targets that must be held at risk in the largest attack options contemplated, plus the size and purpose of the nuclear reserves. Under current policy the United States seeks to hold at risk a large, diverse, and demanding target set—political and military leadership, nuclear forces, other military forces, and war-supporting industry—on the conviction that the Soviets are best deterred when the United States can successfully threaten the assets most dear to Soviet leaders. By adopting a different approach to what is deterring—for example, by threatening to destroy two-fifths of the Soviet population and three-fourths of Soviet industry, as required by U.S. policy during the 1960s; by eschewing or reducing the weight of attack against particular categories of targets, for example, nuclear forces; by reducing the fraction of targets in each category that must be destroyed; or by moderating the level of damage required to those targets—very substantial reductions could be achieved in the number of weapons required to support U.S. strategy. Fifty percent reductions below nominal START levels are certainly not unimaginable. However, any such changes would be a major departure from many years of practice and entrenched belief about the essence of deterrence and would require major revision of the concepts, forces, and plans necessary to achieve deterrence. Still, such changes appear increasingly likely within the context of prospective START II negotiations.

In addition to reducing the number of strategic nuclear weapons, especially those that are most destabilizing, the United States has sought to limit the development and introduction of those new strategic weapons technologies that are viewed as particularly destabilizing. In the START negotiations, the United States has sought to constrain the production, modernization, and flight testing of new types of heavy ICBMs and to prohibit some forms of basing for ballistic missiles, along with a number of lesser restrictions. Otherwise, the United States foresees a continuing need for modern, effective nuclear forces, and the U.S. START position is to allow strategic modernization and the development and deployment of new types and kinds of weapons. The U.S. desire to retain technical weapons options was highlighted by the Senate ratification hearings for the INF treaty, where the effect of that treaty in limiting futuristic delivery systems was vigorously debated before finally being accepted.

STRATEGIC DEFENSE AND SPACE ARMS CONTROL

In the realm of strategic defenses and space, U.S. arms-control policy changed markedly during 1983–1985. To help cement a strategic nuclear relationship based on stable offensive nuclear forces, the United States in 1972 concluded the ABM treaty, which was designed to effectively ban strategic ballistic-missile defense (thereby reducing the value of strategic air defenses as well). American policy toward space arms control was complementary: as far as possible, preserve space as a sanctuary for the operation of those military assets that are needed to make mutual deterrence workable and survivable. Weapons of mass destruction were banned from space by the Outer Space Treaty, and negotiations to ban antisatellite weapons (ASATs) were conducted in the late 1970s.

In 1983, however, the Reagan administration initiated the Strategic Defense Initiative (SDI) program and explicitly challenged the utility and propriety of continuing to rely on deterrence based on offensive nuclear forces alone. The United States sought to develop and, if appropriate, deploy strategic defenses against ballistic missiles, notably to include defenses based in space. Given this approach, the United States entered into the Defense and Space negotiations in Geneva in 1985 with the objectives of protecting the SDI research and development (R&D) program and exploring an alternative arms-control regime. The new regime would supplant the ABM treaty and, building on the offensive force reductions of a START treaty, facilitate a transition to a new form of deterrence based on a growing component of strategic defense coupled with a declining element of strategic offense. The United States pursued a new ASAT research and development program and agreed to consider ASAT arms control only within the context of the D&S negotiations.

Since 1983 these two ideas of how future deterrence might best be achieved have competed against each other in the United States. Neither side has won a decisive victory: the SDI R&D program has proceeded apace, but funding from Congress has been well below that desired by the administration and SDI supporters. Generally, the results of research conducted to date seem to indicate that defenses capable of exerting decisive advantage against very large numbers of offensive missiles are unlikely to be available for deployment until early in the twenty-first century. Accordingly, there are now two basic choices open to the United States regarding strategic defenses. One is to proceed with research on defense technologies through the 1990s (probably at lower than planned levels of funding) in an effort to develop high-leverage defenses, and to decide at that time if deterrence based on a growing element of strategic defense is a desirable, feasible strategic and arms-control option for the United States. The second path is to develop and deploy in the 1990s smaller, less ambitious defenses that would be consistent with the principle of offensive-based deterrence. Such defenses principally would protect the ability of the United States to ensure effective retaliation against a massive attack and might provide quite effective defense against an accidental or small-scale attack but would nonetheless require abandonment or modification of the ABM treaty. For reasons of budgetary stringency, and because of opposition from U.S. allies and the Soviet Union to U.S. strategic defenses in any form, it is most likely that for the next decade the United States will retain the ABM treaty regime and perhaps pursue new space arms-control agreements to protect military assets in space, while pursuing a substantial R&D program on advanced strategic defenses.

THEATER NUCLEAR ARMS NEGOTIATIONS

In the theater nuclear forces (TNF) area, the United States and the USSR in 1987 signed the INF treaty to eliminate ground-based missiles with a range of 500–5,500 kilometers. The elimination process will be completed soon, and unilateral and negotiated reductions of remaining TNF may occur. These developments, along with the dramatic changes under way in the Soviet Union and Eastern European countries, are curtailing the role of nuclear weapons in U.S. and alliance strategy.

The U.S. arms-control position toward TNF differs markedly from START because of the unique political and military circumstances of the North Atlantic Treaty Organization (NATO) alliance and its historic reliance on the doctrine of flexible response. In NATO strategy, as for U.S. strategic nuclear policy, U.S. strategic nuclear weapons fill the role of the ultimate and necessary deterrent. But in NATO the deterrent effect of strategic nuclear weapons was extended to deterrence of attack by conventional, chemical, and TNF of the historically superior Warsaw Pact forces. The primary role for NATO TNF was to provide linkage to the U.S. strategic deterrent by posing a credible threat that NATO would, deliberately or as a result of the momentum of a deteriorating nonnuclear battle, unleash a U.S. strategic nuclear attack against the Soviet homeland.

While attempting to sustain the nuclear forces to support flexible response, the United States for many years pursued a parallel effort to bolster the conventional forces of NATO and to create a better balance between NATO and Warsaw Pact conventional forces. If successful, this would have strengthened the direct defense element of flexible response and reduced the pressure to employ nuclear forces and to threaten nuclear escalation early in a war. This effort was not markedly successful. The same effect was unexpectedly achieved in 1989, however, with the breaking of the Berlin Wall and the collapse of Communist governments in Poland, East Germany, Hungary, and Czechoslovakia. The prospect of a Soviet attack on NATO is now very remote, and the capability of NATO to defeat a conventional attack is now nearly ensured. Accordingly, the need to threaten nuclear escalation to respond to a failing NATO conventional defense has largely disappeared (although NATO nuclear forces are still needed to deter nuclear threats against NATO). NATO's response has been to declare that TNF are now weapons of last resort, and to initiate a far-reaching review of NATO strategy that will, presumably, endorse this less visible and less central role for TNF.

The conclusion of the INF treaty, the prospective conclusion of a START treaty, the demise of the Warsaw Pact, and the major reductions in Soviet conventional forces under a Conventional Forces in Europe (CFE) treaty raise questions about the future of the short-range, remaining intermediate-range, and strategic nuclear forces of the NATO alliance. How can NATO best structure a smaller nuclear arsenal without INF missiles, and with prospective reductions of the remaining TNF arsenal? And how many TNF must be retained to sustain deterrence in these new conditions?

First, NATO must decide what to do about its nuclear requirements below INF range. For many years there has been strong sentiment, especially among the Germans, to reduce or eliminate short-range nuclear weapons such as nuclear artillery shells. In 1990 the United States agreed to forgo modernization of its nuclear artillery in Europe, not to deploy the follow-on to the *Lance* missile to Europe, and to engage in arms-control negotiations to deal with the remaining short-range nuclear forces (SNF) with a range of less than 500 kilometers. There are widespread expectations that SNF will be eliminated from the NATO arsenal.

Second, NATO must decide how to deal with its remaining intermediate-range nuclear forces—aircraft-carried bombs—and the proposal to deploy a new tactical air-to-surface missile, both of which are permitted by the INF treaty. On one side, there is a conviction on the part of the NATO military and many government leaders in the United States and other NATO states that a continuing, effective NATO nuclear force, stationed on the European continent, will remain a critical ingredient of NATO and European security. On the other side is a desire, especially among public groups and some political parties on the Continent to reduce nuclear weapons to an absolute minimum, even to zero, as part of a movement toward greater European peace and political unity. A likely outcome is the removal of all nuclear artillery and short-range missiles, coupled with an agreement to reduce, but retain a limited arsenal of air-delivered weapons, preferably deployed at airbases in several European countries. A fallback could be retaining basing facilities but not actually storing nuclear weapons at these airbases, or even, as a last resort, stationing some number of TNF offshore from the Continent.

A collateral effect of the popular sentiment favoring theater arms control is the possibility of increased opposition to British and French programs for new or modernized nuclear forces, such as a British air-to-surface missile. Whether or not such weapons fall into the INF range, any new nuclear programs are certain to be viewed by some as contrary to the spirit of the INF treaty (to which neither country is a party), arms control, and improved relations with the Soviet Union.

FUTURE NUCLEAR FORCE REDUCTIONS

With the round of nuclear force reductions beyond START, INF, and CFE, many circumstances may converge to relegate U.S. nuclear forces to primarily a retaliatory deterrent role. Such a change probably would be accompanied by a shift to forces designed to be, above all, stable, even at the sacrifice of attributes useful for some types of military missions. The United States and the USSR might both come to favor, or to accept as inevitable, such a diminution in nuclear weapon roles, and if so they might formalize their agreement in arms-control treaties. A change as fundamental as this, however, if undertaken by either the United States or the Soviet Union, would require a major alteration of their present nuclear doctrines, with major implications for their overall military postures and strategies.

The world is taking a shape very different from that we have known since the beginning of the nuclear age. While the United States will remain a predominant world power in economic, political, and military terms, the economic power of the United States relative to that of other advanced states continues to decline. The future of the Soviet Union, on the other hand, is largely unpredictable. The Soviet Union could succeed in domestic reform, becoming either a greater threat to the United States than during the 1970s and 1980s or a more benign and status quo power. Alternatively, the Soviet Union could fail in its reforms and continue on its course as a declining world power, even to the point of dismemberment. Some other nations—the PRC, India, Japan, Brazil, a more united Europe—could (or for good reasons may not) complement their growing economic might with sufficient military power to create a more multipolar world in military terms. In any case, the United States will face many states around the world that are more prosperous and more technologically sophisticated in comparison with itself than is the case today.

In this world the United States could choose to draw back from less important security commitments or could rely more on mutual security arrangements or could assume a power-balancing role among more equal nations and groups of nations. The United States could choose to spend more of its defense resources on conventional, power projection forces that may be in heavy demand in such a global environment. In such a world the U.S. nuclear deterrent would not be irrelevant; in fact, in political as well as military terms it would remain an irreducible element of deterrence against major nuclear threats and could be an even more important guarantor of ultimate American and Western interests against a broader set of contingencies from new quarters.

While the basic deterrent role of U.S. nuclear weapons is unlikely to change in such a world, several circumstances may push the United States to shed some of the less central roles traditionally assigned to its nuclear forces. First, there is little reason now to anticipate that the United States will seek to revive an effective damage-limiting nuclear strategy, or the nuclear superiority that would probably be needed to achieve such a capability, even if that should become technically feasible once again. There is a long-standing trend in the United States to abandon nuclear war-fighting and reject nuclear weapons as legitimate military tools. Rather, nuclear weapons are very widely viewed as useful only for deterrence of the most direct, mortal nuclear threats to U.S. survival. For example, President Reagan denied that his nuclear modernization and SDI programs were designed to achieve nuclear superiority and stated that "a nuclear war should not be fought and cannot be won."

It is also possible that nonnuclear weapons will become capable of assuming at least some of the more discriminating strategic tasks that can currently be accomplished only by nuclear weapons. Although the use of strategic nonnuclear weapons would be unlikely to prevent a major central war from going nuclear, such forces could plausibly be chosen to achieve discrete targeting objectives in limited attack options and could also be used to meet U.S. goals for reduced collateral damage to nonmilitary assets near targets, thus replacing or supplementing nuclear weapons in some roles and missions. Such weapons would almost certainly be chosen in preference to nuclear weapons if long-range forces are needed in conflicts with nonnuclear-weapon states.

Finally, it is possible that defenses will become better able to counter offensive weapons at both the nuclear and nonnuclear levels. This could throw doubt on the effectiveness of smaller scale nuclear attacks, leading to investment in defensive rather than offensive forces with the marginal defense dollar.

If the United States chose or was forced to allow nuclear weapons to recede to a deterrent-only role, fundamental changes in U.S. strategy would be necessary. First, it would be essential to ensure that the basic deterrence of direct, large-scale nuclear attack against the United States and its allies would remain viable. In terms of its nuclear war plan, the United States would probably no longer feel it necessary or feasible to lay plans to attack offensive nuclear forces in hopes of limiting further damage against the United States. Plans could focus instead on attacking other assets judged to be of high value to the Soviet regime and/or most important to destroy on the basis of U.S. calculations of deterrence. These could be soft, fixed area targets, like military-supporting industry, or soft, mobile area targets, like nonnuclear military forces. They could equally well be fixed or mobile point targets, such as Soviet military and political leadership facilities. But in particular, the requirement for prompt destruction of a very large and hardened target set, as is implied by the counterforce mission, and for many attack options against small and diversified target groups, as is implied by flexible response, might be replaced by fewer attack options against smaller groups of assets in a less compressed time frame.

In terms of U.S. strategic forces themselves, such changes in strategy could lead to an even greater emphasis on very survivable nuclear offensive forces, such as SLBMs and mobile ICBMs, and on slower-reacting nuclear forces, such as bombers and cruise missiles. The number of offensive nuclear forces required would probably be considerably smaller than today and would not necessarily change with fluctuations in Soviet nuclear forces. Such reductions are clearly contemplated in the U.S.–Soviet joint statement of objectives for START II negotiations. In the more distant future, however, changes as radical as the reduction to 3,000–4,500 nuclear weapons, or the complete elimination of nuclear ballistic missiles, could even be plausible, insofar as the threat to stability from ballistic missiles might come to outweigh their (declining) ability to limit damage to the United States, and assuming effective survivability and penetration by the remaining nuclear forces. Relying safely on such an attenuated force would undoubtedly call for further steps to ensure the survival of the supporting elements of the offensive forces, including space-based early warning and communications. Strategic defenses could also figure prominently in the strategic force structure, by protecting the U.S. nuclear retaliatory capability, and ensuring that neither countermilitary nor other limited, selective attacks by an enemy would succeed. However, insofar as stability and mutual reassurance remain prominent objectives of U.S. nuclear forces, the combination of offensive and defensive forces would have to be designed so that they clearly would not pose an effective first-strike capability against the Soviet Union.

There are certainly other ways that the future could unfold. Principal among these in terms of impact on U.S. nuclear forces would be the emergence of a military challenge (Soviet or otherwise) large enough to convince the United States to attempt to acquire damage-limiting capabilities, or to restore the credibility of flexible response. However, such a political determination would have to be supported by the development of new technical capabilities, probably both offensive and defensive. While effective damage limitation, or even a convincing ability to engage in and control escalation, appears technically infeasible today, the development of such capabilities, although risky from the point of view of the U.S. objective of ensuring strategic stability, cannot be ruled out indefinitely.

Strategic Nuclear Arms Control

The geopolitical conditions and U.S. national security policies suggested above would be quite amenable to being addressed by arms control. Relaxed demands on nuclear forces could translate into readiness to establish a stable nuclear balance at low levels of nuclear forces, perhaps through negotiated reductions, and perhaps coupled with strategic defense and space arms-control measures. The approach currently favored by the United States is to continue in the direction taken in the START negotiations, by seeking to further improve stability through approaches such as moving the ICBM force entirely to single-warhead mobile missiles, stationing a larger fraction or all of the nuclear ballistic-missile force on submarines, or placing an even larger fraction of the nuclear force on survivable aircraft.

To complement such a move toward crisis-stable deterrent forces, the United States might also seek through arms control to reduce threats to the command, control, and communications for the nuclear forces and to the national command authority (by prohibiting electromagnetic pulse weapons, establishing SSBN keep-out zones, etc.). Such ideas have already been suggested by some arms-control advocates. They could be more acceptable in an environment where nuclear weapons have been placed in a more secondary national security role.

Revolutionary arms-control approaches, such as the total elimination of nuclear weapons suggested by Soviet President Gorbachev, seem largely chimerical even in the distant future. Similarly, the abandonment of nuclear arms control seems plausible only in the presence of such complete relaxation of tensions among nuclear powers that it is not needed, or in the presence of threats of central war so great, and over a protracted period, that the United States would commit itself to a nuclear war-winning strategy. Only the former condition appears even remotely plausible today.

If the United States were in the future to pursue strategic nuclear arms-control negotiations aimed at placing U.S. and Soviet nuclear forces in a deterrent-only role, there would be secondary effects on other elements of U.S. national security policy and on the security of American allies and friends. For one thing, it would force the serious consideration, and probably the inclusion in negotiations, of the nuclear forces of all important nuclear powers (at least France, the United Kingdom, and the PRC). Both the French and the British are currently planning to expand their strategic nuclear forces to several hundred nuclear weapons by the mid to late 1990s, placing them primarily on nuclear submarines with a life expectancy into the 2020s.[2] Reductions far beyond START would bring U.S. and Soviet forces down to the level where the British and French forces, and the forces the PRC might deploy by that time, would be too large to be ignored. Indeed, both the British and French, while affirming that national nuclear forces are essential to their security, have indicated that at significantly reduced levels of U.S. and Soviet strategic nuclear forces (and in the presence of stringent limits on U.S. and Soviet strategic defenses) they would feel obliged to participate in strategic nuclear arms talks.

Should that become the case, the United States would need to develop a concept for determining appropriate numerical limits among several nuclear powers. The United States might enter such five-power negotiations with a relatively low and finite floor for its own nuclear forces; such an approach would be facilitated if the United States abandoned a counterforce targeting strategy.

Alternately, the United States could pursue the idea of establishing relative force sizes as in the Washington Naval Treaty.

Strategic Defense and Space Arms Control

In the early twenty-first century, through the conduct of a dedicated SDI research program, the United States should have assessed the technical balance of advantages between strategic offenses and modern strategic defenses. One of three options is likely to be chosen if defenses appear feasible and advantageous. The United States could propose (for reasons of cost, predictability, etc.) to continue to ban defenses in order to retain a stable nuclear balance based on offenses alone. This would mean retaining, if possible, some version of the ABM treaty, and probably measures to protect space assets that support offensive retaliation. The second course would be for the United States to deploy defenses to reinforce a stable balance of offensive forces (principally by safeguarding the ability of the United States to execute a retaliatory nuclear attack) and to defend against small-scale nuclear attacks. Finally, the United States could deploy very capable defenses to reduce the effectiveness and credibility of offensive forces, and eventually to attempt to supplant offensive retaliation as the basis for strategic deterrence.

If the decision were taken to deploy defenses in any role, the United States would probably seek, as it has done in the D&S negotiations, to use the arms-control process to establish an agreed-upon strategic rationale and legal framework for such a deployment. This could mean seeking Soviet agreement to modify the ABM treaty or the negotiation of a new treaty which would jointly manage the transition by coordinating the deployment of defenses by both sides. The U.S. objective, here again, would be to ensure that strategic stability and crisis stability would be protected as defenses are deployed and offenses are (presumably) reduced. The deployment of defenses could hasten the demise of noncentral nuclear-weapon roles.

The chief alternative for the United States as an objective for deploying very effective strategic defenses is to develop protection against ballistic missiles and aircraft that could be effective enough, in combination with offenses, to offer a realistic damage-limiting strategic force for the United States (although this would probably be possible only if U.S. defenses were far more advanced than those of the adversary). This is one of the few plausible cases where the United States could deliberately choose a course of action designed to ensure that strategic nuclear forces would retain a preeminent role in future U.S. security strategy. The United States has understood from the beginning of the SDI program that defenses could pose this revolutionary capability; however, U.S. administrations have pledged not to use strategic defenses to seek superiority if advanced defenses prove feasible. This disavowal has been warmly endorsed by U.S. allies.

There simply does not appear to be a large constituency in the United States that would favor such a policy, and in any case the technical basis for movement toward such a policy is today only theoretical.

Theater Nuclear Arms Control

One of the conditions that could lead the United States to shift its nuclear forces into a deterrent-only role would be a reduced need to provide extended nuclear deterrence—the nuclear umbrella—to our NATO allies. This is coming to pass both because of the collapse of the Warsaw Pact and because of the growth in power and technical sophistication of our NATO allies.

Improvement of the relative nonnuclear force balance on the European continent, along with a reduction in the tensions that could lead to war in Europe, has long been recognized as the sine qua non of effective NATO deterrence with less reliance on nuclear weapons. Those are the only conditions that would allow for nuclear reductions that go far beyond START and INF, without jeopardizing U.S. and Western security. For these reasons, nonnuclear forces and the conventional force balance are increasingly becoming the centerpiece of U.S. security thinking. NATO has already relegated its nuclear response to a nonnuclear attack to a last-resort role, and substantial reductions beyond the INF treaty are anticipated, both in short-range nuclear forces and the remaining INF weapons. The United States prefers to see the retention of a sizeable NATO TNF arsenal for theater deterrence purposes, which would probably be composed of gravity bombs, stand-off missiles carried by aircraft, and submarine-launched ballistic missiles assigned to NATO by the United States and the United Kingdom. Alternatives are the withdrawal of U.S. nuclear forces to the NATO flanks, probably the United Kingdom, Italy, and Turkey, or from all Western European nations, so that U.S. nuclear weapons offshore (both ballistic missiles and cruise missiles on submarines) and stationed on a permanent basis in the United States provide a less visible theater deterrent for NATO's reduced extended deterrent need. Another option is that a more politically unified Europe could over time come to deploy its own nuclear deterrent force, based on U.K. and French nuclear weapons, with or without some link to U.S. nuclear forces. In each of these cases, the demands on U.S. nuclear forces for the theater deterrent role would be less demanding and less central to U.S. security needs.

CONCLUSIONS

There are many reasons to believe that the United States will seek in the future to place nuclear weapons in a more secondary role than they fill today, so that they continue to provide deterrence against direct, large-scale nuclear attacks against the United States and its allies but are not expected to be the basis for strategies requiring nuclear first use, deliberate escalation, or a carefully controlled application of strategic nuclear force. The United States could deliberately choose to move in this direction to further the long-standing U.S. interest in establishing greater geopolitical and crisis stability. As noted above, many circumstances press for movement in this direction as well.

The United States is almost certain to continue to try to use arms-control negotiations as a preferred method to move in the direction of its chosen national security policy. There is always hope that this approach will reduce the high cost of security, and because arms-control agreements are viewed widely as a way to establish normality—that is, peace—in international relations, it is politically very popular in the United States and among U.S. allies.

However, major reductions or alterations in U.S. and Soviet strategic nuclear forces beyond the present round of START would require fundamental alterations in their defense strategies and nuclear-weapon roles. Before such changes occur, it will be necessary to complete, implement, and digest the INF and START agreements, and to develop a new national strategy, nuclear policy, and strategic offensive and defensive arms-control position. Such fundamental changes in the global environment, technology, and national strategic policies appear increasingly likely. If such dramatic changes do occur, leading to significant revisions in U.S. nuclear forces, it will be over a quite extended period of time, since powerful arguments and strong constituencies would resist any major reduction in nuclear roles and missions.

Finally, movement in the direction of relegating U.S. nuclear forces to a deterrent-only role would fulfill some of the underlying objectives of decades of U.S. and Western arms-control policy. But as such an outcome becomes more plausible, the costs and risks of achieving such a sea change in U.S. national strategy appear more clearly. However, it is not too soon for the United States to begin assessing the actions that may be necessary to accommodate such a change. These actions could include a long-term reshaping and strengthening of U.S. and allied nonnuclear forces and defense strategy, and the integration of nuclear, conventional, and strategic defense arms-control policy (and rhetoric) so that they can assist with so fundamental a strategic reordering.

NOTES AND REFERENCES

1. The U.S. case for sublimits is made in U.S. Arms Control and Disarmament Agency, Office of Public Affairs, *The Case for START Sublimits*, February 24, 1988. The U.S. and Soviet positions on numerical sublimits is in *Nuclear and Space Talks: U.S. and Soviet Proposals*, published periodically by the U.S. Arms Control and Disarmament Agency, Office of Public Affairs. The sides agreed at the Washington summit in December 1987 to specific warhead numbers for

existing ballistic missile systems. See "Joint U.S.–Soviet Summit Statement," *Survival* (May/June 1988), p. 268.
2. UK Ministry of Defence, *British Defence Policy*, May 1989, pp. 16, 18; French Ministry of Defense, *The 1984–1988 French Defense Programme*, pp. 9, 10; see also International Institute for Strategic Studies, *The Military Balance, 1988–1989*, p. 213.

5

Strategic Nuclear Weapons after START

Michael M. May

INTRODUCTION

The Strategic Arms Reduction Talks (START) negotiations, which have been under way since 1982, have resulted in the United States and the Soviet Union agreeing on major issues regarding the final form of a START treaty. In the START negotiations some kinds of strategic weapons, such as bombs and cruise missiles, are not fully counted or are outside the count so that the total of strategic warheads will be substantially higher than 6,000. Thus, a bomber carrying only gravity bombs and short-range missiles will count as one warhead, even though it may carry as many as twenty weapons. A U.S. bomber carrying long-range cruise missiles will be considered to carry ten warheads even though it will carry up to twenty missiles. Sea-launched cruise missiles (SLCMs) will be outside the total of 6,000, and SLCMs with ranges over 600 kilometers will be subject to a politically binding limit of 880.

Some sublimits have been agreed upon: no more than 4,900 ballistic-missile warheads on intercontinental ballistic missiles (ICBMs) and submarine-launched ballistic missiles (SLBMs); no more than 1,540 warheads on heavy ICBMs; and ballistic-missile throw-weight limited to one-half the Soviet capability as of a date to be determined. There will be no limit on the number of mobile ICBM launchers, and there will be a limit of 1,100 warheads for each side on mobile ICBM warheads.

Michael M. May • Lawrence Livermore National Laboratory, Livermore, California 94550.

The complex, basic verification package has yet to be completely agreed to. And there remains an underlying disagreement between the United States and the Soviet Union on how to treat strategic defenses, a point we return to below.

I have four comments regarding the START negotiations. First, a 50 percent cut in the current numbers of weapons is not being contemplated. The cut in actual weapons will be more like 20–30 percent from the present numbers, which are in excess of 11,000. It must be noted, however, that budgetary considerations in both countries may well bring remaining warheads closer to the initially advertised 6,000. For instance, the number of B-2 bombers, which are slated to be the main carriers of the uncounted bombs and short-range missiles, is likely to fall short of the original request for 132 aircraft. It is not clear whether the United States will field any land-mobile ICBMs. The Soviet Union may well cut its nuclear modernization program under the severe stress of its economic problems. It is likely, therefore, that START II negotiations will soon follow the present ones and lead to lower agreed-upon numbers.

Second, a greater proportion of the U.S. forces under START will consist of bombers and sea- and air-launched cruise missiles. As mentioned, this is coming about through U.S. insistence—an insistence due to two factors: (1) the belief that the United States is currently ahead in this technology and (2) a stated preference for air-breathing carriers because of the potentially longer warning time associated with them (i.e., because they travel much more slowly than ballistic missiles).

This second point, however, hinges on the existence of an effective warning system, against cruise missiles in particular. Such warning systems are at present absent or deficient, and they would have to be built from scratch for cruise missiles. In the case of future cruise missiles, this could require major technological advances, with costs running to tens of billions of dollars. Unless such a warning system is built, there will not be any warning at all, as compared to one-half hour with ballistic missiles.

A third comment, often noted, is that this agreement must rely on more intrusive and extensive verification than has been needed for previous agreements, including more extensive on-site verification than has been required for the Intermediate-Range Nuclear Forces (INF) Treaty signed in 1987. Verification of the INF treaty is easier than it will be for START because a total ban on and the destruction of certain kinds of weapons are what must be verified for INF. The START agreement verification provisions, on the other hand, must provide for the counting of numbers of the weapons that are still permitted to be manufactured, changed, and deployed, and that is a much tougher job.

This greater reliance on intrusive and extensive verification measures has both good and bad features. The good features have been prominently discussed; on-site inspection makes for somewhat greater openness between the two countries, particularly in areas about which there has been concern and worry. The bad feature is that, for these measures to be politically viable, there have to be good relations between the two countries. Intensive verification systems are fine as long as Soviet–American relations are good. If relations become tense and unfriendly again, maintaining such verification measures may very well prove to be a politically intolerable burden for one side or the other. They will be expensive, and they will be offensive to some people. All in all, they may amount to a less robust and more fragile verification arrangement than the old-fashioned reliance on national technical means, which are going to be there in good times or bad and provide, in general, strategically and militarily adequate estimates, if not precise counts.

The final comment is that there is a fundamental difference of views between the United States and the Soviet Union on how to handle strategic defenses. The Soviet Union remains in favor of continued strict adherence to the Anti-Ballistic Missile (ABM) Treaty, with no deployment of additional defenses. The United States, at least as of now, is in favor of the deployment of such defenses if they are found to be technically feasible, survivable against attack, and cost-effective. It may be that these differences will be papered over. Such a papering over sometimes emerges in the course of negotiation, and at various times in the past the United States and the Soviet Union have seemed to want them papered over. Yet we are talking about a fundamental difference here.

Clearly the deployment of nationwide defenses would change the impact of various levels of offenses, and it would do so whether the defenses are known to be effective or not because it would alter the entire characteristic of the exchange. The deployment of nationwide defenses would alter the time lines, it would alter the nature and extent of warning potential, it would provide different target sets for the two sides, and it would alter damage expectancies on the two sides for various targets. It would be a great complicating factor. This complication and added uncertainty could be good or it could be bad—it surely would be expensive.

STRATEGIC CONSEQUENCES

What are some of the strategic implications of a proposed START agreement? The work that George Bing, John Steinbruner, and I did[1] focuses on three of these implications for a particular version of the agreement, out of a great many that are possible.

We asked, first, whether the 50 percent cutbacks that were originally contemplated would affect the survivability of the U.S. and Soviet strategic deployments for various force configurations. We also looked at deeper cuts to approximately 25 percent of current forces. As we are all aware, survivability is the sine qua non of stable deterrence: the forces deployed must, at the very least, be capable of surviving an attack. To the extent that they are not, a crisis can precipitate a race to strike first.

Second, we asked what happens after reductions to the ability of the two sides to attack certain target sets, a measure of the military adequacy of the reduced forces. In order to answer that question we postulated some target sets, actually a most important and controversial topic to which I shall return.

Third, we tried in a very preliminary way to look at the impact of the postulated force reductions on the totals of civilian fatalities, which are, of course, the major dimension of nuclear war.

Force Survivability

We looked at force survivability for the present nuclear forces and for forces like them, but cut by a factor of 2, maintaining the proportion of the various force components. We also looked at modernized forces, with the modernization taking the direction of making the forces more survivable—for instance, by putting more submarines at sea, heightening the bomber alert rate, and replacing the silo-based missiles with mobile missiles. We looked at such cases for a total of 6,000 and 3,000 weapons on each side. Even 6,000 weapons is below the numbers that would be permitted under the counting rules for bomber weapons now contemplated in START, as I have noted. For the reduced forces we assumed nuclear weapons on ICBMs, SLBMs, and bombers, the latter carrying cruise missiles but no bombs or short-range missiles.

The key assumptions that were made for the mutual vulnerability assessment are shown in Table 1. We assumed that command, communication, control, and intelligence (C^3I) would work. This is a pivotal assumption; whether it is true or not, however, does not in the first instance depend on the number of weapons. Thus, while the survivability of C^3I is essential for stable deterrence, examining it is a separate matter from examining the impact of deep cuts.

We assumed that one side—the side that strikes first—would be fully generated (i.e., it would have as many as possible of its strategic weapons in an operational state). The side that was struck would be on normal alert status (i.e., it would have taken no special steps to be ready for war). This is a very pessimistic assumption for the side that is being struck. It means (assuming, for instance, that the United States is the side that is being struck) that the United States would either have taken no action during a crisis buildup based on the warning that it might be getting or that it would get no such warning. Most analysts believe that this is a very low-probability situation; it is the situation that displays the maximum vulnerability of the side that is struck.

Table 1 shows our assumed alert rates, both for the side that strikes first and for the other side; the alert rates are conventional ones generally used in such studies. We increased them somewhat for the 3,000-weapons case on the assumption that it would take a long time to reduce to that level and the day-to-day posture at that time would include a higher proportion of total forces. We as-

Table 1. Typical Assumptions on Alert Rates, Reliability, and Survivability for U.S. Forces

Alert Rates

At generated alert:	
ICBMs	100%
SLBMs	70–75%
Bombers	95%
On day-to-day alert:	
ICBMs (in silos)	90%
ICBMs (mobile)	75%
SLBMs	60%
Bombers	30%

Reliability

System reliabilities for ballistic missiles are assumed to be 80%.

Bomber reliabilities, including bomber or ALCM penetration to targets, are assumed to be 70%.

Survivability

ICBMs in silos require two or three attacking warheads for high confidence of kill.

Mobile ICBMs would be deployed to require at least five attacking warheads for each mobile warhead destroyed.

Submarines and their SLBMs are assumed invulnerable.

Alert bombers and tankers will have sufficient warning of missile attack to safely escape and to make barrage attacks impractical.

The strategic C^3I system is sufficiently redundant, hardened, and/or mobile to ensure its function for retaliation.

(Assumptions for Soviet forces are similar.)

---

sumed an overall system reliability of 80 percent. We made penetration assumptions (really cruise-missile penetration assumptions), survivability assumptions that again would be standard for this kind of study. None of the results are terribly sensitive to these assumptions, within reasonable ranges. Thus we assumed that to destroy an ICBM in a silo would require either two or three attacking warheads, depending on the kind of silo. We assumed that to barrage mobile ICBMs effectively would take five warheads for each missile, which makes barraging a very unrewarding tactic. We assumed that alerted bombers would safely escape from base and would therefore have to be barraged if they were going to be attacked at all; and barraging bombers is also very unrewarding. Another pivotal assumption for the United States was that all submarines at sea would remain invulnerable.

The results of force survivability assessments can be displayed by drawdown curves, which display the results of a first strike by one side against the other. Figure 1 shows a drawdown curve for the present forces, at least a stylized version of the present forces, with 10,000 warheads on each side (the general look of the graph would be the same, whether the forces are taken to be 10,000 or 11,000). The two forces start out initially at the upper right-hand corner of the graph at 10,000 each. The curves are an accounting of how many forces are left after each kind of attack.

The case where the Soviet Union makes a first strike against the United States (an assessment of U.S. vulnerability) is shown below the diagonal dashed line. Above the diagonal line is an assessment of Soviet vulnerability under U.S. first strike. The two are almost mirror images of each other, but not quite, because the force configurations are different. To take the case of the United States being struck, U.S. submarines in port and U.S. nonalerted bombers are easy targets to destroy with a very small expenditure of Soviet forces compared to the U.S. forces that would be destroyed. The next U.S. force component to

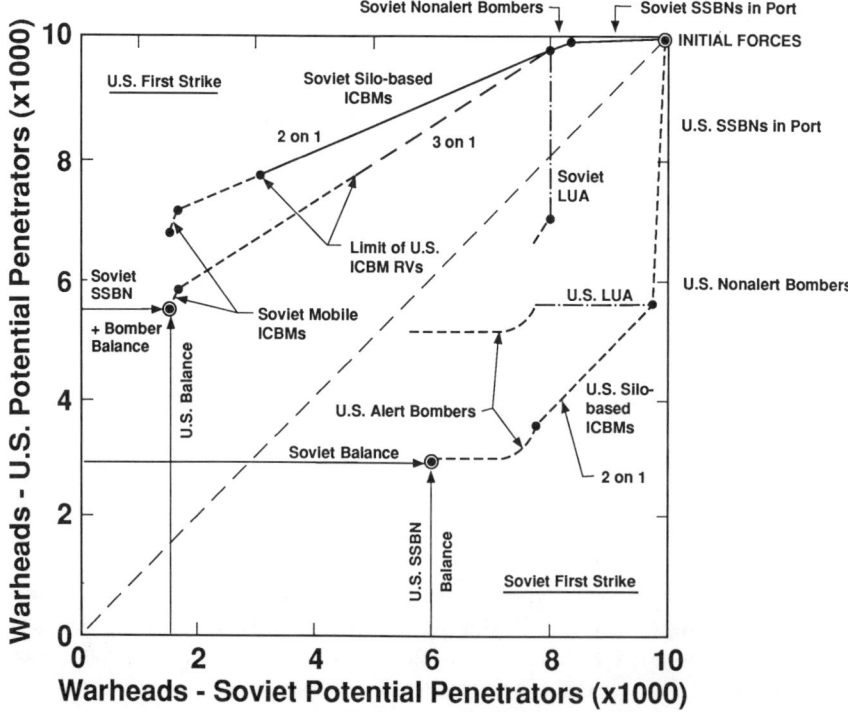

Figure 1. Baseline forces consist of 10,000 warheads for each side and are approximately current forces.

consider is the silo-based ICBMS. The United States has a choice, in principle, of launching the ICBMs under attack or riding out the attack.

If the United States launches under attack, then the survivability curve flattens out, which means that the United States essentially retains all its weapons although it has had to use them in an impulsive response. If the United States does not launch under attack, then the silo-based ICBMs are destroyed with an assumed force-exchange ratio of 2:1, that is, the next solid segment of the curve. The Soviet Union must use a significant number of weapons in order to destroy these U.S. weapons. The next segment is an attack on U.S. alerted bombers by barraging them after they have taken off—a very ineffective tactic because the curve flattens out rapidly because alerted bombers can be only partially barraged.

Following this kind of logic we have the case of a Soviet first strike with the Soviet Union having somewhere between 6,000 and 8,000 weapons, and the United States having somewhere between 3,000 and 4,000 weapons left. Mathematically, this is an advantage for the Soviet Union, but operationally it does not mean much of anything because 3,000 or 4,000 weapons, as will be seen, are enough in our judgment to carry out the most important targeting tasks under present doctrine.

What happens if these forces are reduced to one-half or less? Figure 2 (Case A) shows drawdown curves for reduced forces of 6,000 warheads on each side, and the forces are approximately a proportional reduction from the current force mixtures. Comparison of the drawdown curves with those for 10,000 weapons in Figure 1 suggests a similar pattern, but reduced in scale. After a Soviet first strike the ratio of remaining forces would be similar to the baseline case of Figure 1, but at a lower level.

Figure 3 (Case B) is based on the assumption that the United States and the Soviet Union have modernized their forces in such a way as to make them more survivable, and therefore first strikes are much less effective. Both sides start with 6,000 weapons. In the case of the Soviet first strike, the U.S. submarines in port, the U.S. nonalert bombers, and the nonalert mobile ICBMs are destroyed by a small number of attacking weapons. The nonalert systems are easily destroyed, but the alert systems survive because today there is no practical way to destroy them.

In Case B, Figure 3, if the Soviets chose to barrage the alert U.S. mobile ICBMs and bombers, they would need to use all but 1,000 warheads of their force while the United States would have 1,500 surviving warheads (all on submarines). If the Soviets forwent the inefficient barrage attacks, they would have most of their 6,000-warhead force remaining, but the United States would have about 3,000 surviving warheads. If there had not been any modernization, you would have a somewhat worse situation as illustrated previously for Case A in Figure 2.

Situations with an initial level of 3,000 warheads on each side (which are

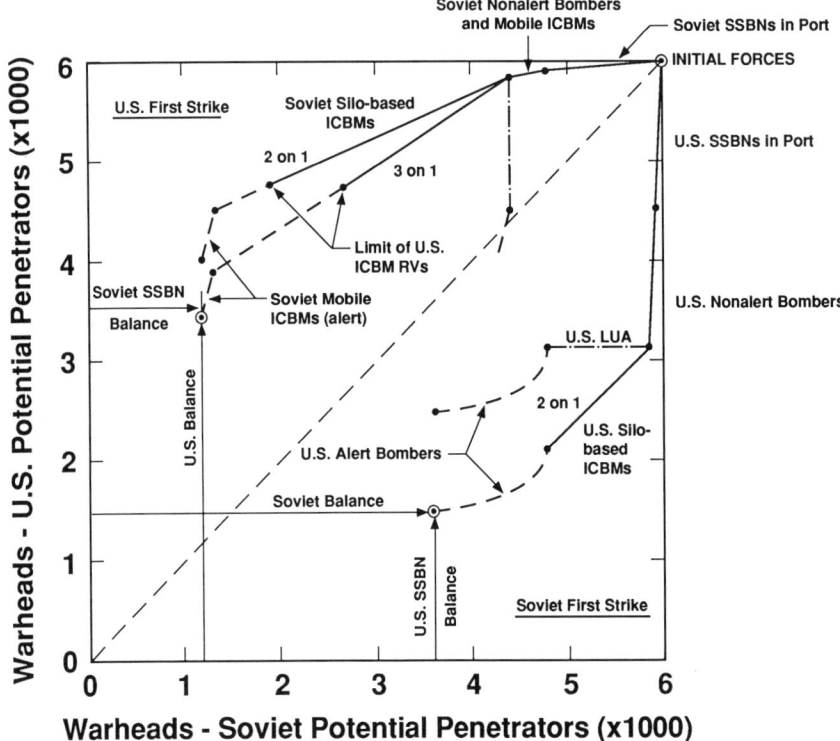

Figure 2. Case A forces consist of 6,000 warheads for each side and are approximately a proportional reduction from current forces but with the B-1 replacing the B-52 bomber.

not illustrated here) exhibit a greater dependence on survivability and alertness. As the number of warheads goes down, survivability and alert rates matter more and more. How much they matter has to be evaluated in the context of what job they have to do.

Target Coverage

A major aspect of our study was the simulation of strategic exchanges that included estimates of the coverage of types of strategic targets in addition to the counterforce targets illustrated in the drawdown curves. We used the well-known Arsenal Exchange Model for our analyses. The model requires numerous inputs describing weapon and target characteristics and a targeting doctrine.

Targeting doctrine is a matter of considerable dispute, debate, and rhetoric, and it has some significant implications for the political aspect of the U.S. strategic posture. One could, at one end of the spectrum of opinion, assume that

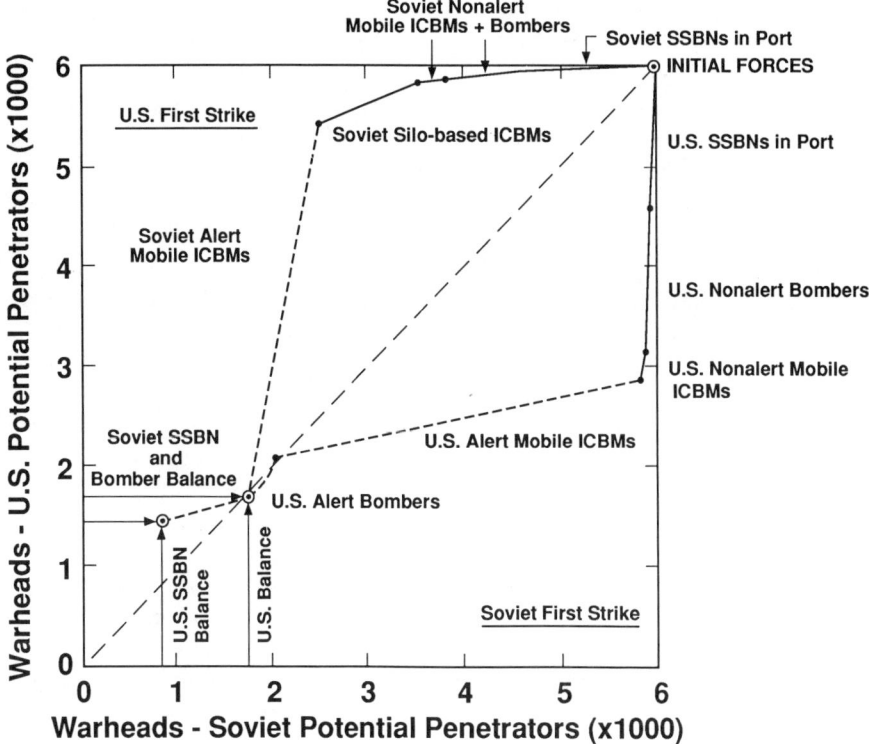

Figure 3. Case B forces consist of 6,000 warheads for each side. The forces are chosen for high survivability and stability. Alert mobile missiles are deployed to force an exchange ratio of 5:1.

100 deliverable nuclear weapons are a good enough deterrent for all possible contingencies, in which case targeting is probably restricted to, at most, 100 centers of industry; or one can go to the other extreme, make a complete list of Soviet military and economic assets (a list that would extend to the tens of thousands), try to determine the top 10,000 or so on that list, and call that the targeting requirement.

In our study we followed present U.S. targeting policy, which identifies four categories: the nuclear forces, the general purpose forces, the military and political leadership, and the economic and military base. Regarding the first category, some of these nuclear forces, the nonalert portion, can be easily destroyed. These make up a large force and presumably are lucrative targets. The alert parts of the force, as noted earlier, cannot be destroyed by current weapons. There is a question associated with this category: Should the United States develop systems that would make the other side's forces more vulnerable than they are now? For instance, should the United States develop systems to go after the Soviet mobile missiles? This is not a settled question. Soviet forces, like American ones, can be drawn down but not eliminated. That is a desirable situation from the standpoint of stability but undesirable if there is a war.

The second category is the other side's nonnuclear military forces, for instance, the Soviet Union's capability for launching an invasion of Europe, including base camps, logistic supply nodes, and so on.

The third category, consisting of leadership and control targets, includes communication, command, and control centers; some space control facilities; and so-called leadership targets. The latter are not a well-defined target set because we don't know what and where their leadership would be in case of war. Some hardened command posts would be in this category.

The fourth category of targets consists of the industrial and economic base, particularly the war-supporting industries. There is, in fact, a large degree of colocation between such defense industry and civilian industries in both the USSR and the United States. There is also a large degree of colocation between these industries and population centers.

In each of these categories, targets do not have the same value, and there is, therefore, a tendency toward a diminishing return with an increasing number of targets, as is exhibited in the next two figures. Figure 4 shows a relative value function plotted versus aim points for Soviet ICBMs. Presumably the Soviet launch-control centers are more valuable than the individual silos because they have some value in controlling more than one silo. The SS-18 silos are more

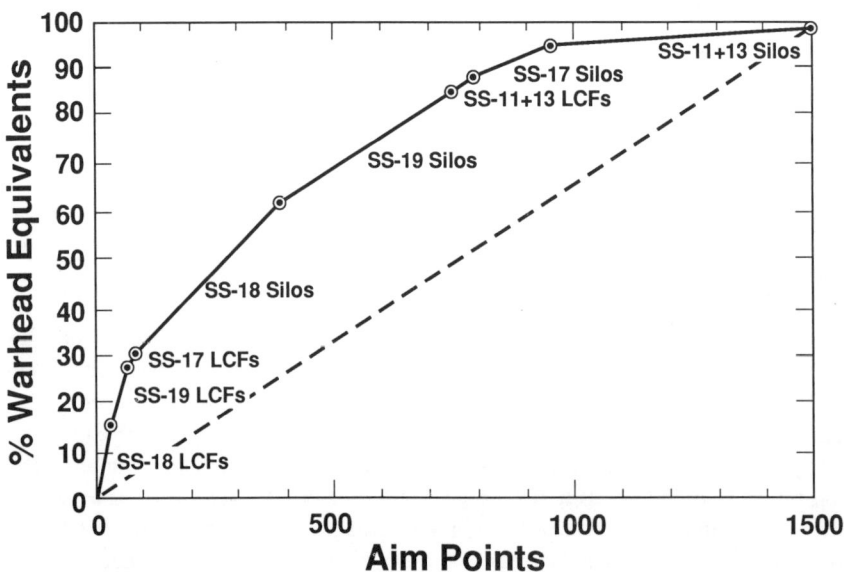

Figure 4. A relative value function for Soviet ICBMs.

valuable than the other silos because they contain more warheads. It is clear that one destroys far more than one-half the total value of these Soviet ICBMs by attacking the first one-third or so of all the silos and control centers.

We can develop similar curves for the other categories. In the case of civilian or industrial assets, illustrated in Figure 5, we see several examples of the same phenomenon. Industrial capacity is shown on the left in two ways. One is the cumulative value or capacity of the top 300 cities. The other is to show the cumulative percentage of industrial installations within 1.5-nautical-mile ranges of the centers of industrial development. In both cases you will see, in different degrees, a pattern by which the first 100 or 200 weapons do proportionally most of the damage. On the right of the figure is a similar curve for the cumulative electrical generating capacity. All these curves thus exhibit a knee. The exact shape of these curves depends on the distribution of value and the degree of colocation of industrial targets, on the yield of the targeting weapons, and on a host of other details, but the general idea is clear. In our study we allocated enough weapons to each target category to work up to the knee of the curve, about 70 percent destruction. In a few cases we destroyed more (99.5 percent in the case of submarine bases), but in most cases the natural stopping place was at about 70 percent.

These are not official U.S. criteria, but one could argue that ours were adequate criteria, at least to develop a preliminary estimate of sufficiency. By that estimate almost all the forces we looked at were sufficient; that is, almost all of them were adequate to cover these target categories.

We assumed an approximately equal total number of targets on the two sides, though the distribution of target types differs. We also took account of the decreasing number of strategic nuclear targets as cutbacks were taken. We did

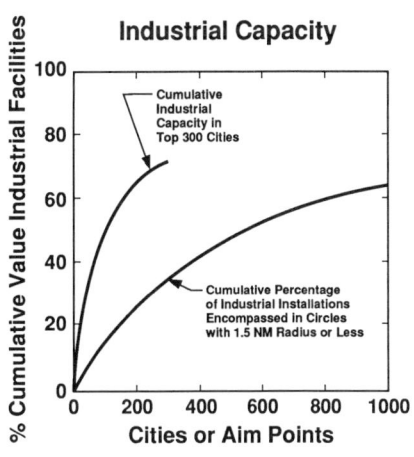
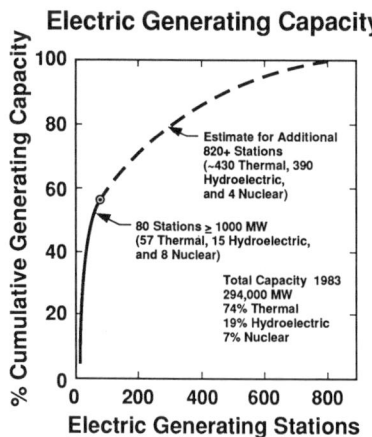

Figure 5. Relative value functions.

not assume any conventional cutbacks, which would have decreased the targets even further. We allowed for some 10 percent of the forces in strategic reserves in each case.

The cases where we did not have sufficiency appeared only when the cuts went down to the 3,000-weapon level, not at 6,000 or 8,000 as we're now contemplating. For a 3,000 level in the unmodernized case we either did not achieve the target coverage desired, or we did not have any weapons left over for reserve, or we did not have either. Again we were not saying that such coverage or reserves are necessary—the study gives the quantitative framework within which one can argue as to what is necessary.

I should note that some of the people professionally involved in targeting would disagree with our assumptions and methods. Many such targeting specialists would regard target arrays in the range of 5,000–8,000 as necessary to carry out the deterrent mission under the present guidance. Many others would consider only a few hundred weapons necessary for deterrence and would judge that going after military targets is unnecessary and harmful. With our assumptions and methods we arrived at a total number of weapons on target of 1,500–2,000, more or less, depending on how far the cutbacks had proceeded.

I want to caution that the target sets we used not only are unofficial sets (we got them by consulting unclassified publications, for instance, geography books) but also that all-out attack against targets in all categories is not necessarily the only, or the most likely, targeting option for the United States or the Soviet Union. What we have developed, as just noted, is simply a metric that permits comparison of the potential performance of forces of different compositions and provides a framework for further discussion.

Civilian Fatalities

A third aspect of our study deals with what happens to fatalities in the case of deep cuts. Briefly put, nothing much happens to them. They continue to range in the tens of millions, whether we have 3,000, 6,000, or 10,000 weapons. The reason is that most of the fatalities occur as a result of the few hundred weapons that are targeted against industries in urban areas. The extent of fatalities depends significantly on the yields of the weapons used against industrial targets because the urban area destroyed along with the industries targeted depends significantly on yield. But going from 10,000 to 3,000 weapons does very little by itself toward reducing fatalities, and the fatalities remain both very high and very uncertain. The assessment we made was quite a simple one, based on the existing target sets; on an assumption of 100 percent lethality at and above 5 psi and zero elsewhere; and on subdividing the two countries into urban, near-urban, and nonurban areas and using approximate population density for each area. Even if a more complex methodology is used, there remains something like a factor of 2 uncertainty on the number of fatalities, owing to such residual uncertainties as the weather and the likelihood and size of fires.

DISCUSSION

In a START agreement we are not contemplating any significant numerical change that would require changing targeting doctrine. If we have the right targeting doctrine now, then it is going to continue to be the right targeting doctrine, and if we have the wrong one, it is going to continue to be the wrong one.

We have to concern ourselves with the survivability of command, control, communication, and warning, just as we did before; this continues to be the top item on the technical agenda if we are concerned about survivability.

After suffering an attack the United States would have most of its residual forces on submarines, and thus the submarine security program will continue to be one of our most important technical programs. But submarines could become endangered without the United States knowing about it, because the survivability of submarines depends on the likelihood that the signal they generate, either acoustic or other, will be lost in the general noise of the ocean. If this ceases to be a correct assumption, we might never know about it. The Soviet Union would not need to deploy additional forces to put our submarines at risk—it could know where they were and go after them when it wanted to with existing aircraft or attack submarines.

There is a premium, therefore, on finding another system that is designed to have the same survivability characteristics as submarines; namely, that it can survive, without a need for warning, against a first strike by the other side. This brings us to the possibility of developing and deploying land-based ICBMs with such characteristics. How to do this is one of the longer lasting debates of the past two decades.

Neither of the systems now being considered for land-based deployment meets the above criterion. Both of them require adequate warning. The hard mobile launcher, the so-called *Midgetman*, requires fifteen to twenty minutes of warning, and the rail-mobile garrison system requires many hours to a day of warning. Many authorities believe that we would have adequate warning; that is, the Soviet Union would not go to war out of the blue, or at the very least Moscow would configure its forces to get ready for war. If Moscow did know how to destroy submarines, that would be a debatable assumption.

Thus, there is a case to be made for a land-mobile missile that could survive attack without warning. There are concepts for such deployment, but they have not to date been politically acceptable. Essentially, they involve the multiple aim-point system in one version or another. This is a continuing concern.

The possibility of Soviet weapons being covertly deployed over and above the number allowed is also a concern. This possibility exists even now, and it will exist at limits of 6,000–8,000 weapons. It is difficult to see what the consequence could be if we have done our technical homework on survivability correctly. If our systems are clearly survivable and not just marginally survivable, then the Soviet Union cannot destroy them, even with several thousand hidden and covertly deployed missiles. The best it can do would be to get an advantage in force exchange by having a larger residual reserve than we do. But, if we have provided for adequate strategic reserves ourselves, it would be hard to see what advantage would accrue to the Soviet Union. That, however, is a question of judgment.

Two other important and controversial questions arise as we envisage strategic force reductions. One has to do with the desirability of switching the forces toward air-breathers, that is, cruise missiles and aircraft. As I mentioned, these systems are stabilizing only if we actually deploy an expensive and technically advanced warning and defense system—at the very least a warning system. The alert fraction of air-breathing systems is lower than the alert fraction of submarines or land-mobile systems. For that reason, the United States would have to deploy more in order to have the same number survive an attack. Such air-breathers have penetration rates probably lower than those of ballistic missiles. In general, they have lower reliability and entail some targeting inefficiencies. In order for cruise missiles, and especially bombers, to penetrate, some weapons have to be devoted to the suppression of air defenses. And because bombers have limited routing possibilities, they cannot hit an arbitrary set of targets as readily as can missiles. (Highly MIRVed—multiple, independently targetable reentry vehicles—missiles have similar inefficiencies in hitting targets. This is an argument against them as deep cuts proceed, quite aside from the problems about their survivability.)

Bombers are very popular with the Air Force, and so far they have had a greater bureaucratic weight behind them than have survivable missile systems. Whether or not that should continue to be the case is a major question.

We should also revisit the targeting debate. Two questions are sometimes considered together, but they are separable. First, what is the role of counter-military targeting? The doctrine we have now is one stating that there should be no usable force exchange advantage. Is this the right targeting doctrine? I believe it is. One can argue that it is not necessary to aim at military targets, and that doing so will in fact multiply the number of weapons we need from a few hundred to a few thousand, thereby increasing the scope of the arms race. This is a debate we have been engaged in for some time, a debate that clearly will be revisited.

The second question concerns the extent to which we can maintain extended deterrence, to include our allies within the deterrent umbrella, as we cut back on strategic weapons. Answering this question involves assessing perceptual and political factors rather than the kind of factors in our studies, especially for reductions to a level such as 6,000 or 8,000 weapons. Extended deterrence depends mainly on keeping NATO as a full partner in all nuclear deployment and arms-control decisions. At whatever levels of nuclear and conventional forces are eventually reached, NATO must remain the main military factor in Western Europe, integrating the Western European countries, providing them with deterrence, and offsetting whatever Soviet force projection capability remains. The demise of NATO and the consequent American disengagement from Europe would risk the revival of potentially dangerous conflicts and uncertainties among the Europeans themselves—no matter how attenuated the Soviet threat might become.

NOTES AND REFERENCES

1. Michael M. May, George F. Bing, and John D. Steinbruner, "Strategic Arsenals after START," *International Security* (Summer 1988), pp. 90–133.

6

Strategic Arms Control and American Security
Not What the Strategists Had in Mind

Michael Nacht

In the winter of 1986, Thomas Schelling, the intellectual father of modern arms control, lamented that "arms control has certainly gone off the tracks. For several years what are called arms negotiations have been mostly a public exchange of accusations; and it often looks as if it is the arms negotiations that are driving the arms race."[1] When Schelling wrote these words, there was certainly a good deal to be skeptical about. Not a single significant arms-control accord had entered into force between the United States and the Soviet Union since the ratification of the Strategic Arms Limitation Talks agreements signed in May 1972 (SALT I). Subsequent agreements reached between President Richard Nixon and General Secretary Leonid Brezhnev to implement a threshold test ban on nuclear weapons and to restrict peaceful nuclear explosions were never ratified by the U.S. Senate. The superpowers had reached agreement on a statement concerning the prevention of nuclear war and had modified the protocol to the Anti-Ballistic Missile (ABM) Treaty. And they both continued to adhere to most of the provisions of the SALT II treaty, which was signed in 1979, although it, too, was never ratified by the Senate. By the winter of 1986 the prospects for future negotiated agreements seemed slim indeed. Both the Strategic Arms Reduction Talks (START) and Intermediate-Range Nuclear Forces (INF) negotiations appeared completely deadlocked, acrimony dominated the Soviet–American dia-

---

Michael Nacht • School of Public Affairs, University of Maryland, College Park, Maryland 20742.

logue, and the superpower arms competition was as intense as it had ever been in the four decades of the Cold War.

Less than two years later, however, in December 1987, the Soviet Union and the United States completed a far-reaching agreement calling for the elimination of all intermediate-range nuclear forces. This disarmament agreement requires the Soviet Union to destroy 851 launchers and 1,836 missiles and the United States to dismantle and destroy 283 deployed and nondeployed launchers and 867 missiles. These measures are to be completed by the end of 1991. Moreover, the treaty calls for verification measures of unprecedented intrusiveness to ensure that during the decade to follow none of these weapon systems are reactivated.

This notable accomplishment of the Reagan administration has, in due course, been followed up by President George Bush. As the Bush administration was completing its first year, prospects for negotiated arms-control agreements seemed reasonably bright both with respect to START and the newly launched Conventional Forces in Europe (CFE) talks, which are aimed at effecting major, asymmetric reductions of North Atlantic Treaty Organization (NATO) and Warsaw Pact conventional forces deployed in Central Europe.

What has produced this extraordinary turn of events? Was Schelling simply misguided in his pessimism of 1986? Is this progress all due to the flexibility of one man—Mikhail Gorbachev? Indeed, what progress has been accomplished by the INF treaty? These questions steer us toward the more fundamental issues: What are the objectives of arms control? What are we seeking to achieve? How do arms-control agreements relate to the security of the United States, its allies, and the foreign-policy objectives of the Soviet Union? Only if we seek to understand this last set of issues are the more recent developments explicable.

WHAT IS ARMS CONTROL ALL ABOUT AND WHO SAYS SO?

The concept of disarmament, in which active military forces are decommissioned or destroyed, has a long and undistinguished history, marked by utopian thinking and failed international dialogues such as the Hague Conferences of 1899 and 1907. Arms limitation or control, by which is meant the placing of quantitative ceilings on the number and type of deployed forces, is a newer concept that was implemented with limited but only temporary success in the Washington Naval Treaty of 1922 and the London Naval Treaty of 1930. In the case of the former, Britain, the United States, Japan, France, and Italy declared a ten-year moratorium on capital ship construction, agreed to limits on permitted total tonnage of this class of ships, and limited the size and armament of these vessels. In the case of the latter, Britain, the United States, and Japan extended the moratorium until 1936, constrained the maximum displacement of their destroyers and submarines, and set limits on total tonnages for destroyers, submarines, and cruisers. Yet five years later all these major powers were at war.

It was only in the 1950s that sober, strategic analysis was applied to the objectives of arms control. Following the lead of President Dwight Eisenhower's intriguing but failed attempts to achieve a U.S.–Soviet open skies agreement in 1955, the scholarly community took a hard look at arms control as expressive of limited modes of cooperation between potential adversaries. In the early 1960s, as Schelling himself reflected in his 1986 article, three volumes were published that have stood the test of time and remain the landmark publications in the field: Schelling and Morton Halperin's *Strategy and Arms Control*,[2] derived from a series of summer discussions in Cambridge, Massachusetts, in 1960; Hedley Bull's *The Control of the Arms Race*,[3] based on a series of papers he wrote for a study group of the International Institute for Strategic Studies in 1959–1960; and to a lesser extent, Donald Brennan's edited book, *Arms Control, Disarmament and National Security*,[4] based on a 1960 conference of experts.

In particular, the objectives cited by Schelling and Halperin have long been repeated, and I have termed them the classical objectives of arms control: to reduce the probability of war, to reduce the damage should war occur, and to reduce the resources devoted to the preparation for war. In short, Schelling and Halperin envisioned a multiplicity of steps, some large and some small, that rivals in the nuclear age would be willing to take in order to avoid war and, thus, their own destruction. The authors' reasoning, stemming as much from abstract notions of game theory as from the real world of international politics, suggested that the leadership of both the United States and the Soviet Union, despite their fundamental differences, had in common the mutual objective of survival in the nuclear age. From this realization it was at least plausible to argue that both sides would be willing to take steps that would help themselves even if they also helped their rival. While today this may seem to border on the trivial, it was not the case in 1960, when many, at least in the West, believed that the U.S.–Soviet confrontation was a zero-sum game in which a gain for either side was automatically a loss for the other. Schelling and Halperin sought to demonstrate that there were opportunities, albeit small and difficult ones, for positive-sum moves in which both Moscow and Washington stood to gain. Reducing the vulnerability of their respective retaliatory forces to bolster deterrence would be such a step. This also meant that arms control was not constrained to the limitation of armaments per se. Rather, it was concerned with introducing elements of stability into the strategic relationship, where stability meant reducing the incentives, by whatever means, to initiate nuclear war.

The strategic logic of Schelling and Halperin's arguments was then, and is now, unassailable. If one assumes that governmental leaders behave rationally (an assumption doubted by some scholars but accepted by most in the business of making policy), Schelling and Halperin pointed the way in which the two nuclear adversaries could reduce the risk of premeditated or accidental nuclear war. This logic was adopted, to at least some degree, by the Kennedy and Johnson administrations. Building on a modest success of the Eisenhower administration—the Antarctic Treaty, which was signed in 1959 and entered into force in 1961, prohibited deployment of nuclear weapons in the region and called for on-site inspection—the Kennedy administration completed the Hot Line agreement in 1963, which enhanced communication between Moscow and Washington, and the Limited Test Ban Treaty (LTBT) the same year, which banned nuclear-weapons tests everywhere but underground. The Johnson administration completed agreements with the Soviet Union that prohibited the deployment of nuclear weapons in outer space and, subject to certain conditions that have not subsequently been met, in Latin America. Moreover, the Johnson administration, working with its counterparts in Moscow and London, ultimately concluded the Treaty on the Nonproliferation of Nuclear Weapons (NPT), which was signed in 1968 and entered into force in 1970.

On the surface this appears to be an admirable record by two successive administrations of implementing the teachings of Schelling and Halperin. Closer inspection, however, leads to a more cynical assessment. The Antarctic Treaty signed in 1959 was an outgrowth of activities from the International Geophysical Year of 1957 in which both Washington and Moscow sought to position themselves as champions of world peace in the eyes of international public opinion. An agreement was reached to prohibit the deployment of nuclear weapons in a region of the globe in which neither side's military leadership thought it necessary to deploy such forces. In sum, from the national perspectives of Washington and Moscow, adhering to the Antarctic Treaty had marginal political payoffs and gave away nothing militarily. It did, however, establish the important precedent of on-site inspection.

The Kennedy administration, filled with significant representatives of the Cambridge intellectual elite, made little additional progress until the final months (as it turned out) of President John Kennedy's tenure. The Hot Line agreement made sense because it enhanced communication between the principal adversaries, might presumably be useful in defusing a crisis, had high-profile political benefits, hurt no one, and threatened no one's vital interests. Still, coming only eight months after the Cuban missile crisis, this was no simple agreement to reach. The LTBT, however, was as much a reaction to domestic political pressure in the United States as it was a reflection of strategic logic. The plain truth is that in 1963 President Kennedy was under pressure at home to reduce fallout from atmospheric nuclear tests that placed deadly strontium 90 in mothers' milk. Fortunately for both sides, the American and Soviet technical communities had by this time developed high confidence in their abilities to enhance the capabilities of their nuclear weapons by testing programs underground. Neither side needed any longer to test in the atmosphere, and it was, therefore, not that difficult to relinquish this option given the marginal political gains. Still, it took adroit negotiating by Averill Harriman to complete the agreement in 1963.

The Johnson administration as well was motivated in its arms-control behavior by considerations remote from the reasoning of Schelling and Halperin. The Outer Space Treaty of 1967 was similar to the Antarctic Treaty in that it was an exclusionary agreement prohibiting the deployment of nuclear weapons in a physical zone that the military leadership of the day found uninteresting or unattainable. The Treaty for the Prohibition of Nuclear Weapons in Latin America, or the Treaty of Tlatelolco, established conditions that were difficult to meet, including ratification by Argentina and Cuba, and was of limited strategic significance given the range of modern ballistic-missile submarines. The treaty was of value as a nuclear nonproliferation measure against third parties without significantly constraining the military effectiveness of the superpowers. Similarly, the NPT sought to limit the spread of nuclear weapons to other states. This treaty, a stabilizing measure in American and Soviet eyes, and, I would argue, for the planet as a whole, is nonetheless discriminatory as viewed from New Delhi or Buenos Aires. It also flies in the face of the argument that nuclear proliferation at the regional level could promote regional stability much as it has brought stability to the superpower competition.

In sum, through most of the 1960s the considerable progress in completing negotiated arms-control agreements bore only limited resemblance to the strategic logic of Schelling and Halperin. The one case where the connection was more tightly linked was the aborted SALT negotiations stimulated by Robert McNamara, who, later portrayed in popular accounts as the quintessential rational man who went wherever his logic and his data led him, was, in fact, persuaded by the Schelling logic, especially with regard to the danger of defensive deployments. McNamara, it should be recalled, lectured Prime Minister Aleksei Kosygin at great length at the Glassboro, New Jersey, summit meeting of 1967 on the dangers of missile defense. Defenses were unable to cope with a sophisticated offensive attack, McNamara argued. Their deployment by one superpower would only convince the other that the deployer was planning to strike first and then use its defenses to limit the damage from a retaliatory attack. Defenses were therefore destabilizing—they increased the incentive to strike first in times of crisis. They should be mutually avoided through negotiated agreement. Now here was the logic of Schelling and Halperin applied at the highest levels. If McNamara had remained in office and such an agreement had been signed, we could point to it as a major achievement of strategic logic. But McNamara was forced out of office by President Lyndon Johnson over Vietnam policy differences, and efforts to get SALT started in the summer of 1968 were abandoned in the wake of the Soviet invasion of Czechoslovakia. So the application of strategic theory to arms-control policy never materialized.

To be sure, an ABM treaty was concluded under the Nixon administration and with it an interim agreement to limit strategic offensive forces. But these accords, comprising the SALT I agreements, had less to do with strategy than with geopolitics and domestic politics. When President Nixon entered office in January 1969 with his newly appointed national security advisor Henry Kissinger at his side, a new team entered the White House that, over time, saw arms control both in a broader geostrategic perspective and in a more narrow domestic political perspective than dictated by the rational imperatives of Schelling and Halperin. The Nixon–Kissinger priorities were to extricate the United States militarily from Vietnam without losing Vietnam to the Communists, to take advantage of the Sino-Soviet split to influence the Soviet Union, to use arms control as a wedge to blunt Soviet foreign-policy objectives, and to serve Richard Nixon's reelection plans for 1972. With 20-20 hindsight, we now know that only some of these objectives, all in fact interconnected, were achieved.

At the heart of the Nixon–Kissinger plan was to provide incentives for the Brezhnev Politburo to adopt policies less hostile to Western interests. For the United States to get out of Vietnam, seen by Nixon and Kissinger as a gigantic waste of American resources and attention, required Soviet assistance both diplomatically and, more importantly, in reducing the flow of Soviet arms to the North Vietnamese. But how could the United States gain leverage over Soviet policy? What were our carrots and sticks? One stick was to move closer to Beijing and exploit the Sino-Soviet confrontation. That Russian and Chinese troops had fought on the Sino-Soviet border in 1969 and that Ambassador Vladimir Semyonov, the Russian SALT delegate, approached Gerard Smith, the head of the U.S. delegation, about possible joint actions against the Chinese left no doubt in Washington that the conflict between the two Communist giants was real and that American movement toward a less hostile relationship with the People's Republic could possibly exert influence on Soviet policies.

The carrots had to do with arms control but were not of the sort envisaged by the strategic logicians (although Halperin had worked in the McNamara Pentagon and drafted the early concept papers on SALT and then served on the Kissinger National Security Council staff until he resigned over Vietnam policy differences). The reasoning followed from the following question: How could the United States convince members of a hermetically sealed decision-making process with an ideology anathema to the West to adopt policies not hostile to the West? The answer that Nixon and Kissinger came to was that Russians and Soviets, like all humans, are acquisitive beings who want the better things in life once they know such things are attainable. What was needed was a wedge into the Soviet Union so that the society could be exposed to Western goods and ideas. Once so exposed, the people would demand a better life, which, in the short run, could only come from Western imports. In short, we had to hook the Russians on the same materialistic possessions that we ourselves have long been accustomed to acquiring. But how could Soviet society be so penetrated? Only if there was a considerable warming in the superpower political relationship would economic relations begin to improve. And how could this political warming be achieved?

"Through arms control" was the Nixon–Kissinger answer. The principals in the White House believed that the number and lethality of the nuclear arsenals possessed by both Washington and Moscow were so enormous that quantitative restrictions of one sort or another, while of great significance to the defense specialist, were, in fact, meaningless in military and political terms. Therefore, why not go for a major arms-control agreement with the Soviet Union, even if we sacrificed in certain technical areas, when the result could be a wedge into the opening of Soviet society? Moreover, to guard against military uncertainties from Moscow and to sell the agreements at home to the generals, to Senator Henry Jackson, and to other skeptics, Nixon and Kissinger made sure that the United States held onto two trump cards: the freedom to deploy multiple, independently targetable reentry vehicles (MIRVs), which would greatly increase the lethality of the U.S. land-based and (eventually) sea-based forces, and to continue to strive for greater missile accuracy to enhance the hard-target kill capability of U.S. forces. Kissinger argued in closed session to Jackson's committee that these two capabilities ensured U.S. nuclear superiority even while the interim agreement granted the Soviets temporary quantitative superiority in terms of the numbers of launchers. The SALT I agreements, in other words, bought some marginal strategic stability, as McNamara and Schelling believed, but the real purpose was not to reduce the likelihood of war but to seduce the Soviet leadership into a set of international relationships that would ultimately limit its ability to wage an aggressive foreign policy antithetical to Western interests. For, it was argued, arms-control agreements would create a political climate that would lead to greater economic, cultural, and scholarly intercourse that would, in time, open up Soviet society. The gains to the Soviet Union from these interactions would be considerable, and the leadership would then have much to lose by engaging in belligerent actions against Western interests, actions that would put such Western assistance at risk. This, then, was the path to leverage over the Soviet leadership. Arms control was merely a tool in a complex game of international power politics. Given the enormous destructive potential of the U.S. nuclear arsenal, even in the wake of negotiated agreements, arms control, reasoned Nixon and Kissinger, was about politics, not about strategic logic.

For Nixon, arms control was also about domestic politics, specifically presidential politics. Given the Watergate revelations, we now know just how concerned Nixon was about his reelection prospects. It is no accident, as our Communist friends would say, that the Nixon visit to China and the Moscow summit to sign SALT I were timed for the spring of the 1972 presidential election year. So SALT, in particular, was designed and timed to serve presidential interests, not merely U.S. strategic interests. There is, of course, nothing wrong with this. President Jimmy Carter subsequently demonstrated the difficulties a president runs into when presidential politics are not taken into policymaking calculations. But it is certainly the case that some of the weaknesses and ambiguities of SALT I, especially the vague definitions of the interim agreement, were papered over by President Nixon in the interests of an agreement that served his reelection plans.

Alas, Nixon and Kissinger were hoist on their own petard. Nixon, of course, fell victim to Watergate and was never able to be effective during his abbreviated second term. In his zeal to be reelected, he won the battle but lost the war. Kissinger was severely hamstrung as a result. The Russians turned Kissinger's logic against him. Instead of Moscow being restrained from acting aggressively for fear of losing the fruits of what came to be known as détente, it was the Americans who embraced timidity. In the wake of SALT I, many Americans took Nixon at his word that the nature of the U.S.–Soviet relationship was changing from one of confrontation to one of cooperation. Consequently, defense spending could be reduced; American vigilance against the Soviet threat could relax. The humiliating American departure from Vietnam, symbolized by the final helicopter lifting off the roof of the U.S. Embassy in Saigon in April 1975, suggested to many Americans that we had meddled too much and too far from our core interests and had ripped up our society in the process. A period of American retrenchment was at hand.

But this period of retrenchment, we can now see, was matched by a period of Soviet aggressiveness. While Gorbachev may claim today that Brezhnev presided over an era of stagnation, he supervised considerable gains for Soviet foreign policy. New classes of strategic and theater nuclear weapons were deployed in the 1970s that effectively wiped out any semblance of U.S. strategic nuclear superiority. Gains in Yemen and Angola, Soviet belligerence during the 1973 Arab–Israeli war, and the rise of Eurocommunist movements in Western Europe all threatened or had the potential to threaten Western interests. It was the Americans, despite the warnings of an increasingly frustrated Secretary of State Kissinger, who clung to the vestiges of détente and failed to respond. Contrary to the Nixon–Kissinger calculations, and with the help of the Jackson–Vanik amendment, which linked U.S. most-favored-nation status for the Soviet Union to Jewish emigration from Russia, the Soviets insulated themselves effectively from the penetration strategy of America's détente policy. Economic assistance to the Soviet Union from the West was limited and was channeled into the military sector and to the service of Soviet heavy industry. Soviet consumers were told to be patient, as they had been for decades. Soviet political leaders were not seduced. Western leverage over the Politburo was minimal.

The Carter years and the tortuous efforts to achieve a SALT II treaty only continued the process of a deteriorating U.S.–Soviet relationship that began in Nixon's second term. By now the Soviets could mount a serious threat to the entire U.S. land-based missile force, bases of the Strategic Air Command, and submarines in port. Paul Nitze, who resigned in 1974 from the SALT II delegation, championed the concern over this scenario. Nitze, deeply hurt that he was not asked to join the Carter administration, launched a potent and effective attack, first against Carter's choice as head of the U.S. Arms Control and Disarmament Agency, Paul Warnke, from which Warnke never fully recovered in terms of congressional support, and later against the SALT II treaty itself. Nitze and his colleagues on the Committee for the Present Danger, including Eugene Rostow, Richard Pipes, and Kenneth Adelman, convinced many Americans and many senators that SALT II was a bad deal for the United States They argued that it did not constrain the Soviet threat to the American strategic forces; it granted superiority to the Soviets in missile throw-weight and other important quantitative categories; it did not satisfactorily constrain the Soviet *Backfire* bomber; it placed an unnecessary and unwise moratorium on the deployment of ground-launched cruise missiles and intermediate-range ballistic missiles in Europe to counter the newly emerging Soviet threat embodied in their SS-20 force of highly accurate, MIRVed, mobile missiles; and many of its important provisions could not be verified with high confidence.

SALT was on the ropes in terms of strategic logic. But it was rendered defenseless by Soviet actions elsewhere. Allegations of Soviet violations of SALT I could not all be easily refuted. The Soviets followed up their actions in Yemen and Angola with gains in Ethiopia and Nicaragua, and then with the invasion of Afghanistan in December 1979. It was simply too much to ask the American people and their representatives to support a strategic nuclear arms-control treaty that many claimed codified growing Soviet capabilities, or the administration that authored it, in the face of such dissonant developments. Add to this the humiliating experience of the American hostages in Iran, following the fall of the shah and the rise of Ayatollah Ruhollah Khomeini, the establishment of the Sandinistas in Nicaragua, the spread of Vietnamese Communist influence throughout former French Indochina, and even false alarms such as the episode of the Soviet combat brigade in Cuba, which surfaced during the SALT II ratification hearings, and it is easy to see what fueled the rise of a potent conservative political movement in the United States. President Carter would have had to be masterful to navigate among these minefields, but he was not. Given his propensity for bad luck (e.g., the oil price rises and the poorly executed Iranian hostage rescue mission) and the disastrous levels of unemployment and inflation at the time of the 1980 presidential election, his fate was sealed.

It is understandable, therefore, that an alternative view running counter to Schelling's logic and Kissinger's geopolitics would surround arms control in the 1980s. Now Eugene Rostow, Kenneth Adelman, and Richard Perle, Henry Jackson's protégé, as well as others of similar views, were in the Reagan administration making policy. Their view was contemptuous of Schelling's and Kissinger's perspectives. They argued that arms control was like a narcotic for the Western democracies. It lulled us into a false sense of security. It lowered our guard. It led us to delude ourselves that peace was at hand. Arms control was the modern-day umbrella of the Neville Chamberlains of our time. While liberals were concerned about a repeat of 1914 and a nuclear world spinning out of control to a disaster that nobody wanted and nobody planned, conservatives saw the relevant historical analogy as 1938. The big threat, they argued, was not accidental war but appeasement leading to war. The Soviets were Hitler's successors. Arms-control agreements that reduced the West's willingness to spend the necessary funds for its own defenses were only leading us to disaster against an opponent that was plotting to wage nuclear war and to destroy our values and our civilization. Ronald Reagan was elected president to see that this foolishness would stop and that the United States would get its strategic house in order before it was too late.

PERSONALITIES, DOMESTIC POLITICS, AND THE SENSE OF HISTORY

How then can one explain that, under President Reagan and with Richard Perle as a major architect, the INF agreement was achieved? Again, the answer is not to be found in strategic logic, in the language of the defense specialist, in the calculations of the counter-military potential calculus. It is to be found in personalities and politics. Reagan and his colleagues, first of all, found it easier to be in opposition on these issues than to be in the driver's seat. Loose talk by Secretary of State Alexander Haig about a nuclear warning shot across the bow and deeply hostile pronouncements by Reagan toward the Soviet Union created the nuclear freeze movement that spread across the United States and led to demonstrations from London to Bonn to Tokyo that were the largest antinuclear demonstrations since 1945. This political pressure initially forced Reagan to send someone to the bargaining table—General Edward Rowny for the START negotiations and Paul Nitze for the INF negotiations. Whereas Rowny was skeptical of any agreement that could be negotiated, Nitze wanted the job only if he could work toward an agreement. In this objective Nitze had support, for a time, from Richard Burt, who served as director of the State Department's Bureau of Politico-Military Affairs and then as Assistant Secretary of State for European Affairs before being pushed out of the way as U.S. ambassador to Bonn in the second Reagan term. More importantly, Nitze had the sustained support of Secretary of State George Shultz, who, over time, tended to dominate Reagan's foreign and national security policy. Pitted against Nitze was Secretary of Defense Caspar Weinberger; Kenneth Adelman, who replaced Eugene Rostow as director of the U.S. Arms Control and Disarmament Agency; William Casey, director of the Central Intelligence Agency and a close friend of the president; General Rowny, who opposed not only a START agreement but an INF agreement as well; and key defense department aides, especially Richard Perle. While Perle was able to thwart successfully Nitze's efforts to break a negotiating logjam with his famous walk in the woods with his counterpart, Ambassador Yuli Kvitzinsky, Perle turned out to be too clever by half in inventing a formula that seemed equitable on the surface but was thought to be nonnegotiable: the double-zero option, in which both the United States and the Soviet Union would dismantle all their intermediate nuclear forces. Given the considerable numerical superiority held by the Soviets in this class of weaponry, who would have thought that the Russians would agree to such asymmetrical reductions?

What Perle with his plans and Schelling with his logic and his pessimism could not count on, and what, quite frankly, few others could see as well, was the enormous significance of Nancy Reagan and Mikhail Gorbachev in the arms-control process. Mrs. Reagan, always a close advisor to the president, began to believe during the second Reagan term that the president had established his credentials as a tough guy. He had restored American self-confidence. He had generated support to build up America's defenses. His Scowcroft Commission had helped buy fifty MX missiles and discredited the window of vulnerability of U.S. strategic forces even though, in fact, the forces were more vulnerable to a sophisticated Soviet attack than they had been during the Carter years. He had pushed through the NATO INF modernization deployments in 1984. He had scored small gains against tyranny in Grenada; attacked Muammar Khadhafi in Libya; backed anti-Communist forces in Nicaragua, Afghanistan, and elsewhere; and championed the Strategic Defense Initiative as a way out of the box of the terror of nuclear deterrence. Now it was time to be the great peacemaker, to think of the history books, to conclude with a major arms-reduction agreement with the Soviets. An enormously popular president from the far right with this record, even more than Nixon, had little to fear in domestic political terms from striking out on this fresh approach as his endgame strategy. After the deaths, first of Brezhnev, then Yuri Andropov, and then Konstantin Chernenko, he found his opposite number in Gorbachev.

By 1986, shortly after Schelling wrote, Reagan was looking for an arms-control agreement as much for himself and for history as for strategy and geopolitics. Gorbachev, quicker than almost anyone believed possible, came to similar conclusions for fundamentally different reasons. The system of government that Gorbachev came to lead in 1985 was on a need-to-know basis. Probably, until he became general secretary in March of that year, much of the inner workings of the economic situation at home and the national security situation abroad were not known to him. There would be no reason to keep the Politburo member in charge of agriculture so informed.

Once at the top he learned the full truth. He was appalled at what he had inherited. America, the great adversary, had as allies Western Europe, Japan, and China, while his associates were Cuba, Angola, Vietnam, and the basket cases of Eastern Europe. The Soviet army was bogged down in a land war in Afghanistan that had no plausible end in sight. The economy was in a shambles. Alcoholism was rampant in the work force. The old aphorism was true: The government pretended to pay the workers and the workers pretended to work! Creativity and initiative were stifled. The information age threatened to leave the Soviet Union behind as a third-rate power by the twenty-first century (even after four years of *perestroika* there are estimates of only 100,000 personal computers in the Soviet Union compared to 30 million in the United States). Gorbachev clearly needed breathing space abroad to rejuvenate the economy and society at home. He needed improved political relations with the West to gain its economic assistance. Lo and behold! Gorbachev had taken a play out of the Nixon–Kissinger playbook. Fourteen years after Washington had tried détente to penetrate Soviet society, Gorbachev was willing to try the same gamble in order to revitalize his country. Contrary to all past expert predictions, the Soviet system had indeed produced a risk-prone, high-stakes roller willing to preside even over the loosening of the Soviet empire in order to place his country in a stronger long-term position. For Gorbachev, as for Nixon and Kissinger before him, arms control was a means to this end. Did he really need nuclear superiority in the European theater when he wanted support from the West? Wouldn't Soviet-based systems do the job just as well, if necessary? Wouldn't a major arms-control agreement in Europe loosen enormously the cohesion of the Atlantic alliance? Almost assuredly it would. Wouldn't a major diplomatic peace offensive, summitry, withdrawal of Soviet forces from Afghanistan, opening of the Soviet political system, and even opening of Soviet military facilities to intrusive verification and inspection build enormous trust in Western circles?

It is highly unlikely that Gorbachev thought all this through in advance. He clearly has been feeling his way and, in the process, has miscalculated on several fronts. He has underestimated the intense nationality problems within the Soviet Union, including the frustration of leaders in the *Russian* SFSR, led now by Boris Yeltsin, and their willingness to abandon even the party if that is required for survival. He has overestimated the cohesion of the Baltic and other non-Slavic republics within the Soviet Union. He has underestimated the speed with which the rejection of Communist party rule would come in Eastern Europe; and perhaps most tellingly, he has grossly underestimated what it would take to jump-start the Soviet economy. But he has been on target in assessing the receptivity of the West toward arms-control agreements. Whether he will be able to manage these conflicting objectives within his own political structure is anybody's guess. Can he keep the military, the KGB, and nationalist sentiments in line? Can he protect himself against political deposition or assassination? Can he avoid having to make a tough choice between military intervention in Eastern Europe or the Baltics and the complete political independence—perhaps neutrality—of these states? Can he avoid the disintegration of the fabric of Soviet society into full-scale civil war? Can he finally get perestroika to pay economic dividends? How long will the leadership that he has reshaped, the elites, and the masses wait for results? On the answers to these questions turn the future of U.S.–Soviet relations, the future of Germany, and the future of NATO.

Students and opponents of arms control, as talented and as energetic as they are, too often forget the forest for the trees. Arms control is not influenced by politics. It is about politics. It is politics, both domestic and international. This is not to say that there are no objective military conditions. There are. But in the absence of war, only beliefs about military conditions matter. As with deterrence, the only time you know if you have enough is when conflict ensues. When it does you know that deterrence has failed. When it does you find out if you have enough defense and offense. In the absence of war, however, arms races and arms control serve political masters who are seeking to retain political control at home and pursue geostrategic objectives abroad. Arms-control negotiations, their arcane language, and their subtle concepts are tools for these larger purposes. Strategic analysts often like to say: "I want to deal with the substance, not the process. The issues are what matter to me." But the process and the political purposes behind it are as much substance as the technicalities of the agreements themselves. Maintaining such false distinctions may be nice for the classroom but can be deadly in the crucible of the policymaking process.

Hedley Bull put it correctly many years ago:

> Unless the powers concerned want a system of arms control; unless there is a measure of political détente among them sufficient to allow of such a system; unless they are prepared to accept the military situation among them which the arms control system legitimizes and preserves, and can agree and remain agreed about what this situation will be, there can be little place for arms control.[5]

We are still a long way from meeting all of Bull's conditions. We are getting there, more because of domestic demand than because of strategic necessity. Barring fundamental technological breakthroughs comparable in magnitude to the nuclear-tipped, multiple-warhead, highly accurate ballistic missile that would give one side or the other or a third party a striking military advantage, arms control will continue to play a significant role in the domestic politics of the superpowers, a substantial role in alliance politics, and a marginal role in affecting the military capabilities of the superpowers. It is a necessary but far from a sufficient condition for a peaceful transition to the next century. It is not what the strategists had in mind.

NOTES AND REFERENCES

1. Thomas C. Schelling, "What Went Wrong with Arms Control?" *Foreign Affairs* 64 (Winter 1985/86), p. 219. This article was adapted from a presentation at the Nobel Symposium of 1985 on *The Study of War and Peace—Perspectives on Present Knowledge and Research*.

2. Thomas C. Schelling and Morton H. Halperin, *Strategy and Arms Control* (New York: Twentieth Century Fund, 1961).
3. Hedley Bull, *The Control of the Arms Race* (New York: Frederick A. Praeger, 1961).
4. Donald G. Brennan, ed., *Arms Control, Disarmament, and National Security* (New York: George Braziller, Inc., 1961).
5. Bull, *Control of the Arms Race*, p. 10.

7

Beyond German Unification
The West's Strategic and Arms-Control Policies[1]

Lynn E. Davis

INTRODUCTION

The West has historically struggled to combine coherent military strategies with consistent arms-control policies. The political imperatives of arms control have often taken precedence over clarity in strategic thinking, most starkly in the Intermediate-Range Nuclear Forces (INF) Treaty. The excuse for not worrying in the past was the stability derived from the ideological and military division of East and West. After all, military strategies could be changed only on the margin, and arms control offered only limited prospects for moderating the political and military confrontation. Now, political pressures arising from developments in the Soviet Union and Eastern Europe are overwhelming both military and arms-control policies. All the uncertainties are providing the excuse for failing to set strategic and arms-control goals.

The problem is that Western governments are acceding to public demands for reductions in conventional forces without having defined what strategic purposes those forces might need to serve in the future. They are now planning to move immediately to follow-on negotiations after signing the Conventional Forces in Europe (CFE) agreement in late 1990, without having considered what their objectives will be. They have accelerated negotiations on U.S. nuclear

---

Lynn E. Davis • The Paul H. Nitze School of Advanced International Studies, The Johns Hopkins University, Washington, D.C. 20036.

forces in Europe without determining what role, if any, they might play in the future in promoting peace and stability. Obviously, strategic and arms-control policies cannot be divorced from politics, and indeed they must serve the West's broader political goals. But all the lessons of history suggest that a failure to think clearly about the West's strategic requirements and to integrate these with its arms-control objectives holds real dangers.

The first task for the West will be to translate its political goals for further improvements in East–West relations and for freedom and democracy into specific policies and institutional structures for the future of Europe. The 1987 Single European Act, with its promise of a true common internal market in 1992, has relaunched the process of political and economic integration in Western Europe. Notwithstanding the opening of the East, the European Economic Community (EEC) now seems ready to move beyond economic and monetary union to expanded political cooperation among its twelve members.

The issue of the future role of the United States in Europe must also be faced. So far the thrust of American policy has been to find ways to stay. Americans are actively encouraging further economic integration in Western Europe and have called for formal institutional and consultative links with the EEC. They are leading the search for new political and economic roles for the North Atlantic Treaty Organization (NATO), suggesting that it now foster democracy in Eastern Europe and serve as an instrument for the resolution of regional conflict. Yet isolationism in America has strong roots and certainly has not lost its basic appeal.

All of these choices will be importantly affected by what happens in Germany. Moving beyond their past rhetoric calling for the end of the division of Europe, Western governments now hail the reunification of Germany, having achieved their conditions. Germany will be a member of both the NATO alliance and the EEC.

In designing its future military strategies and arms-control policies, the West must also consider the environment in which military forces might be used. As the actual military threats have decreased, the uncertainties surrounding the future bases of security have nevertheless increased. The West no longer fears a Warsaw Pact attack in Europe. This is the case even though large numbers of highly capable military forces remain in Eastern Europe. Those who bear responsibility for military planning will find it difficult to reconcile these divergent political and military realities. Their consolation should be that a Soviet political decision for war in Europe would take considerable time to execute militarily, thereby providing the West with warning and time to respond. The Soviet Union will still have the capability for a limited military attack with little preparation. But it is difficult to posit a political objective that could be served by such an action.

Significant reductions are already occurring in the military forces of the Soviet Union and the countries of Eastern Europe, and more are on the horizon as the CFE treaty is implemented, as the Soviet Union withdraws entirely from Eastern Europe, and as scarce public resources are shifted away from defense. Mikhail Gorbachev has outlined follow-on phases for CFE in which the conventional force postures will be cut even further and restructured defensively. He also has a plan for eliminating all nuclear weapons.

The Soviet Union will remain a superpower, with its arsenal of nuclear weapons, and the predominant military power in Europe—that is, unless the empire were to fracture in the course of domestic reform or under pressures from the different nationalities. Moreover, military conflicts could still arise in Europe as a result of political instability, boundary disputes, and perhaps even by accident or miscalculation. As these are most likely to occur in Eastern Europe, the West's interests will probably not be directly involved. So the serious military threats that the West now confronts are only future possibilities—most important, a resurgence of the Soviet Union as an expansive military power, which could occur if Gorbachev is replaced by reactionary leaders or, especially, if his domestic reforms succeed. There is also the hypothetical threat of the emergence of a powerful and expansive German state.

In spite of—indeed, because of—these political and military uncertainties, the West must define its military strategies and arms-control policies. The CFE and Strategic Arms Reduction Talks (START) agreements, with the establishment of parity in strategic nuclear and conventional forces, provide the point of departure. What follows is a consideration of the West's critical choices for conventional and then for nuclear forces in Europe, where the objectives of East and West have converged on the desirability of significant reductions in conventional arms but still diverge over the future role of nuclear weapons. (Throughout this chapter the terms *West* and *East* are used as a convenience in presenting the issues.) Individual countries will have different perspectives and interests when it comes to specific choices. And over time the future of Europe could well involve very different political and military alignments than those in place today.

CONVENTIONAL FORCES IN EUROPE

The major accomplishment of the CFE agreement, when implemented, will be the establishment of a conventional military balance in Europe. The treaty will promote stability by eliminating the East's capability for surprise attack and large-scale offensive operations. The ceilings to be placed on Soviet forces will have the important effect of requiring the Soviet leadership to violate the CFE treaty in order to mount in the future any credible attack against Germany. The agreement also includes significant measures to verify the CFE levels and could be expanded to monitor any future unilateral reductions. The agreement can also be expected to act as a catalyst to further political and military changes, even though it will take some years for the actual reductions in military forces to occur. At the same time, pressures for further reductions are increasing in the United States and throughout most of Europe. The CFE signing is being viewed politically as an initial phase of an ongoing process of negotiations and reductions.

Under these circumstances the West will confront the immediate issues of planning for its conventional military forces under the CFE ceilings and allocating resources for defense. The Western allies will also need to determine soon their future objectives in conventional arms control. Indeed they shall have to decide whether the arms-control process remains useful or has been overtaken by political events.

What Goal for Conventional Defense?

The CFE agreement establishes ceilings just below current NATO levels on the major items of conventional equipment. The immediate task will be to allocate the reductions among the sixteen NATO nations. Political as well as military considerations will be involved. For future planning, however, the individual contributions under the ceilings are not as important as whether the CFE ceilings will be maintained. For that decision will have the effect of establishing how the West views its goal for a conventional defense.

If the CFE ceilings are maintained, NATO will be in a position, even under the most conservative of assumptions, to mount a quite robust forward conventional defense with the capability to deny the Soviet Union its military objectives in any potential conventional war. Such a defense would hedge against uncertainties as to future Soviet military policies and restructuring. It would permit the West to reduce significantly its reliance on nuclear weapons. It would also be consistent with far fewer U.S. nuclear weapons in Europe.

As an alternative, the West could be satisfied with a more limited goal for a conventional defense, given political changes in the East, and with a significant reduction in the military forces of the Soviet Union and the Eastern European countries. It could, for example, maintain the readiness for a conventional defense similar to what NATO could accomplish today (from seven to fourteen days), or even less—a very minimum conventional deterrent. Such a choice of a goal for a conventional defense would be based on the view that the threat of war in Europe has essentially disappeared. In the future the West's security would not depend upon maintaining forces equal to those of the East as a whole or capable of defeating them under the hypothetical possibility of an attack. All that would be necessary is for conventional forces to be able to hedge against the possibility of war through miscalculation. Nuclear weapons would provide the ultimate deterrent.

The West's choice for a defense will not, however, be determined by highly abstract strategic considerations; rather, the decision will be made as the West chooses whether or not to maintain the ceilings mandated by the CFE agreement. That choice will, in turn, be affected by how the West views the future role of arms control and the importance of mutually negotiated reductions. Maintaining the ceilings could provide useful leverage for negotiations, while failure to do so could obviate any controlled process of mutual reduction.

Public pressures will also come into play, which in this case are somewhat competing. Maintaining the CFE ceilings would require Western governments to maintain defense spending at roughly current levels. But it would permit a significant reduction in nuclear weapons. A more limited goal for a conventional defense would require fewer conventional forces than those mandated by the CFE ceilings and, therefore, less defense spending. But nuclear weapons would remain the primary hedge against uncertainties and the possibility of war in Europe.

Thus budgetary pressures will compete with pressures for reducing reliance on nuclear weapons, at least as long as the West posits a military threat from the Soviet Union and remains committed to the main tenets of its current strategy for a conventional defense.

Defensive Strategies and the Future of Forward Defense

With constraints on both defense spending and military manpower, Western governments have already begun to adopt some of the ideas of those calling for a military restructuring of forces to emphasize defense, especially the proposals to increase reliance on reserves and on high-technology, precision-guided weapons. With the withdrawal of Soviet military forces from Eastern Europe, NATO has decided to move away from forward defense and toward a reduced forward presence.

What this will mean still requires definition. But with the increase in warning of an attack from the East, the West could reduce the number of active divisions and rely more on reserves, decrease the number of military forces forward deployed in Germany, and relax the requirement for when reinforcements from the United States and the United Kingdom would need to arrive. With the establishment of a conventional military balance, the West would no longer need to offset Soviet quantitative superiority with its advantage in technology, although the Supreme Allied Commander, Europe (SACEUR) now argues that smaller forces will be more dependent upon high-technology systems, and that these systems will be needed to offset the possibility of improvements in the quality of Soviet forces.

With CFE and the proclaimed defensive restructuring of Soviet forces, pressures will arise for NATO to remove those nominally offensive elements in its strategy, and especially its plans for deep conventional strikes. To break the momentum of an enemy attack as well as to win with smaller forces, NATO has defined a theater concept, the Follow-on-Forces Attack, to strike successive echelons of Soviet forces with aircraft and missiles. The U.S. Army also has a doctrine for linking close-in and deep battle areas, emphasizing maneuver, counterattack, and strikes into rear areas. To implement these concepts, the United States is developing a reconnaissance strike complex, which involves highly sophisticated intelligence, communications, and weapons. The Soviet Union is clearly targeting these concepts and systems in its call for more emphasis on defensive doctrines.

Concepts of deep strike have in the past been criticized in the West for diverting resources away from ensuring a forward defense and for relying on long-range missiles, which are capable of surprise attack and could carry nuclear as well as conventional warheads. Revising such concepts could save money because their associated weapon systems are expensive and rely on very high-technology sensors. The problem is that the sensors, which provide warning of preparations of an attack, could be critical to the West's defense if reliance is placed more on reserves and reinforcements. The capability for deep strikes could become even more important militarily if the Soviet Union under the CFE agreement deploys most of its counteroffensive potential back in the Western Military District. The ability to hold Soviet forces at risk could also be a deterrent to their movement forward in a crisis.

None of these possible changes in the West's future conventional force posture in Europe promises any major reduction in defense spending. The critical strategic choice for the West is whether under CFE it wishes to adopt a more defensive strategy, clearly forgoing a forward defense and changing more fundamentally the manner in which it is currently implemented.

A variety of different concepts for clearly defensive, or nonprovocative, strategies have been defined in the West over the past decade. Their underlying principle is to design a force posture with the structural inability to launch an attack. They tend to emphasize territorial defense and reliance on infantry militia with high-technology weapons and sensors. They call for static defenses (barriers and mines) and strategies of attrition rather than maneuver. The purpose of the militia forces would be to inflict delays and casualties and to disrupt the attacking force. The proposed strategies place less emphasis on the more clearly offensive forces, such as tanks, strike aircraft, and tactical ballistic missiles, as well as nuclear weapons.

The significant reduction in Soviet armored forces alone makes defensive strategies more attractive militarily. But such strategies have important limitations, primarily their inability to recover lost ground from the attacker or to punish its territory. In reducing the options of the defender, they simplify the task of the attacker, permitting concentration against selected points to make limited territorial gains. Defensive strategies are also very dependent upon trained manpower and on expensive high-technology weapon systems.

If the West were prepared to forgo a forward defense, or change the manner of its implementation, it could adopt elements of such defensive strategies, relying on more mobile operations, giving ground to gain time to discover the enemy's main attack and wearing the enemy down with disproportionate attrition on unfriendly territory, while holding a mobile reserve for counterattack. Such a strategy would emphasize new advanced technologies, including intelligence-gathering sensors, secure communication equipment, surface-to-air defensive weapons, and longer range precision-guided munitions. The military forces as well as the costs of such a defensive strategy could be significantly less than those today. By adopting a more clearly defensive strategy, the West could cut its military forces and defense spending and reduce its current reliance on nuclear weapons, while maintaining a quite robust conventional defense as a prudent hedge against future uncertainties.

The Future Role of American Troops in Europe

The CFE agreement also raises the issue of the future peacetime presence of U.S. troops in Europe. President George Bush in his 1990 State of the Union address proposed that the United States reduce its forces in Europe by 80,000 men from the current level of some 305,000. The Soviet Union initially agreed to this overall level of 225,000 men, with the proviso that only 195,000 men could be deployed in the Central Region of Europe. But as a result of its planned withdrawal of military forces from Eastern Europe, the Soviet Union no longer wished to include limits in CFE on military manpower. The West agreed.

The presence of American troops in Europe today has a military rationale and provides a symbol of the U.S. commitment to the defense of Western Europe. An American peacetime presence in Europe will be significantly less important militarily with the substantial withdrawal of Soviet forces from Eastern Europe. Yet the military reasons for maintaining that presence will not entirely disappear because of geography and the difficulties in reinforcing Europe from the United States. The symbolic argument for maintaining American troops will remain and could become more important if the number of U.S. nuclear weapons in Europe is significantly reduced. But with the political changes and the establishment of a conventional balance, other powerful arguments will be added to those already made by Americans for withdrawing, namely, the pressures on the U.S. defense budget, the ability of Europeans to undertake a larger share of the burden of their defense, and the need to protect American interests worldwide.

The West Germans were ambivalent about the presence of American troops even before the opening to the East. Whether most in Germany will continue to see American soldiers as contributing to the security of Germany is uncertain. Their support for large numbers and for the activities associated with the American presence, particularly low-level flights and military exercises, has unquestionably diminished.

In fact, the number of American troops in Europe has fluctuated quite dramatically over the years. Their size has not been as important as the fact of their presence. Some reductions in American forces forward deployed in Europe could be made simply on military grounds, given the diminished threat and the time it would take the East to mobilize for an attack.

The initial Bush initiative in CFE equated Soviet and American conventional forces in Europe, singling them out for equal limits. One way to proceed in setting the level of American forces in the future would be to maintain equality with the Soviet Union. The problem is that Soviet and American troops in Europe serve very different political purposes, and the Soviet Union will continue to have the geographic advantage of being able to reposition its forces quickly in Central Europe. Another approach would be to set the level of American troops so as to redistribute the burden of defense to the Europeans. Americans could, for example, be given preference in reductions arising from arms-control or a defensive restructuring.

The United States in the past has seen the presence of American troops as contributing to international stability and to its own security. It has sought to leave the Soviet Union in no doubt as to its commitment to the defense of Europe and to the credibility of its nuclear guarantee. Its preference historically has been to keep the nuclear threshold high through a robust conventional defense. In the future the issue will be whether there is a minimum number of troops that the United States will wish to keep in Europe to ensure its own national interests, even in the event of radical political change in Europe and the virtual dissolution of the military threat.

Future Directions in Arms Control

The West will need to maintain forces essentially at the levels established by the CFE agreement if its goals and strategy for a conventional defense are to remain the same as in the past. If the countries were to be satisfied with a more limited conventional defense and/or were prepared to adopt a more defensive strategy, it would be possible to make significant reductions, with the size depending upon which changes were made. The issue then would be whether to make these reductions unilaterally (so as to respond to political and budgetary pressures) or to seek mutual reductions in follow-on negotiations in CFE.

Gorbachev has called for further significant reductions in conventional forces in Europe and for the ultimate restructuring of the armed forces of the two sides for strictly defensive purposes. He has also urged negotiations and mutual reductions when responding to pressures from the Eastern European countries for the withdrawal of Soviet forces. The West agreed to move quickly to follow-on negotiations, with the primary goal of achieving limits on military manpower.

An alternative would be for the West to promote stability in Europe through the establishment of a situation in which no military offensive would have any prospect of success. Such a mutual defensive restructuring of the force postures would remove the uncertainties associated with what the Soviet Union is currently undertaking unilaterally. It would make it easier for the West to adopt its own more defensive strategy. In this case, the West would remain at roughly the post-CFE levels until the more fundamental goal of a defensive restructuring of the conventional force postures can be negotiated.

What such restructured force postures would entail still needs to be defined. It is not so simple as removing all offensive armaments, for weapons systems and technologies will be offensive or defensive depending upon how they are employed. Further significant reductions would be necessary, but they would not in themselves produce this result. According to Klaus Wittmann, the West would need to pursue the related objectives of the redeployment of Soviet formations to the interior of the USSR, a reduction in their state of readiness and of their stocks of combat supplies and material, as well as measures to guarantee that forces in the forward area cannot be rapidly built up.[2]

Negotiations on military doctrine might be a useful first step, but confidence- and security-building measures (CSBMs) would provide the actual means of ensuring that a defensive restructuring had occurred. These measures, by improving information on the military forces of the two sides—their deployment, readiness, training, and peacetime activities—would demonstrate their defensive characteristics and disposition. CSBMs could also be designed to constrain the actual military activities that would be necessary to launch an attack, overtly or covertly. Constraint measures might include limits on the size, location, and duration of mobilization and out-of-garrison activities as well as on the movement forward into Central Europe of military forces, bridging equipment, and ammunition.

The Soviets have agreed to exclude naval forces in the CFE agreement but have informally linked negotiations on these forces to further reductions in Western forces and to their defensive restructuring. Western navies have objected to any limits on their forces or activities, both to maintain flexibility in their operations and because of their peacetime as well as wartime missions. Naval forces would certainly complicate any future negotiations. But given their prominent role in the West's defense, it is difficult to see how they can be excluded. CSBMs would be the logical first step.

A more critical issue for the West will be whether to proceed in further reductions in conventional forces within the limited geographical area of the Atlantic to the Urals. Western Europeans have insisted upon a right of withdrawal from the CFE treaty if the Soviet Union builds up its ground and air forces east of the Urals. A future negotiation will run up against the minimum capability required by Europeans for protecting their own global security interests. The next phase of conventional force negotiations will need to address in some form the total force postures of the superpowers.

In pursuing CFE, the West gave priority to reductions and the defensive restructuring of conventional forces in Europe within the framework of bloc-to-bloc negotiations. But political developments in the East are leading to an end to the division of Europe and to the dissolution of the Warsaw Pact alliance. As a result, continuing the CFE process may not be the best way to proceed in conventional arms control in Europe. The alternative is to give priority to the Conference on Security and Cooperation in Europe (CSCE) and the CSBM negotiations currently under way.

The attraction of CSCE is that it links political and economic issues with security in Europe and involves all the countries of Europe and the United States and Canada. Negotiations occur among thirty-five sovereign nations, with the alliances playing an important but informal role. Following on from the Stockholm Document in 1986, the CSBM negotiations focus on regulating military activities rather than reducing military forces. Their goal is to promote confidence and stability within the whole of Europe. If the CSBM negotiations were to replace a follow-on to CFE, individual governments could proceed to reduce and restructure unilaterally their military forces in light of changing political and military developments, with CSBMs providing the means collectively to observe and monitor the changes. The notification measures in the Stockholm Document could be improved to cover more military activities, including alerts and mobilization. Measures could also be designed placing constraints on the most threatening activities, including movements of military forces out of garrison. Of course, a follow-on to CFE could proceed in parallel with the CSBM negotiations, as is occurring today. The risk is that, absent a choice, neither will be successful, given their competing assumptions as to the political future of Europe.

NUCLEAR WEAPONS IN EUROPE

Arms control will play a central role in the West's future nuclear strategy in Europe. In the spring of 1990 NATO called for negotiations on short-range nuclear forces (SNF) once a CFE agreement is concluded. The problem is that the West lacks a strategic concept to guide these negotiations.

Choices for Strategies and Force Postures

President Ronald Reagan and Secretary Gorbachev's visions of a world without nuclear weapons aside, agreement seems to exist in the West that nuclear weapons are a fact of life and, with varying degrees of confidence, that they contribute to peace and stability. Experts appreciate the insoluble dilemma posed by nuclear weapons. A credible deterrent depends on having effective military, or war-fighting, capabilities. But because of the destructiveness of nuclear weapons, they cannot be used for the classic military objectives of victory in war. If deterrence fails, they also cannot compensate for conventional inferiority, for analysis shows that the use of nuclear weapons, if reciprocated by the other side, will not improve the conventional military situation. Nuclear weapons are maintained so as to prevent war and, if deterrence fails, would be used for the political purpose of convincing an enemy of the unacceptable costs of continuing to use its military forces.

With the implementation of the CFE treaty, NATO will lose what has been its primary public argument for maintaining nuclear weapons—to compensate for its inferiority in conventional forces. With the prospect of a successful defense, NATO will be relieved of having to use its nuclear weapons first in a conventional war in Europe. No longer will there be the requirement for nuclear weapons to be able to deny the Soviets their battlefield objectives through the capability to attack troop concentrations and command centers. As a result, most of the military arguments for short-range nuclear systems will disappear as will the reasons for a large U.S. nuclear stockpile in Europe.

In the past, U.S. nuclear weapons served other important roles: deterring their use by others, hedging against the possibility of a miscalculation as to the nature of the conventional military balance, and demonstrating the commitment of the United States to the security of Europe. What the West needs to decide is whether U.S. nuclear weapons in the future will still be needed for any of these roles. If these were also to disappear with the political evolution of Europe, would there be a residual role for U.S. nuclear weapons in promoting stability and preventing war in Europe?

If U.S. nuclear weapons are judged to be necessary for any of these roles, for it is a matter of judgment and not analysis, it will then be necessary for the West to define their future tasks. One view of the role of nuclear weapons, known as existential deterrence, is that they promote deterrence simply by their existence. In this view, nuclear weapons are weapons of last resort. There is no requirement for their being able to carry out particular military tasks or for specific numbers or kinds of nuclear-weapon systems. SACEUR, in contrast, has based his military requirements in the past on the need for a seamless web of nuclear capabilities, deployed throughout the potential battlefield, with the tasks to destroy those Warsaw Pact military forces posing a threat to NATO in limited strikes (initial and retaliatory) as well as in an all-out retaliation (known as general nuclear response). Designated in SACEUR's current threat assessment are several thousand mobile and fixed military targets in Eastern Europe and the Soviet Union. The implementation of the CFE treaty, as well as the unilateral reductions and restructuring on the part of the East, would substantially reduce that number of targets, but not, in this second view, the tasks for U.S. nuclear weapons.

A third view is that with fundamental political change in Europe as well as the establishment of a conventional balance the actual tasks for nuclear weapons will be significantly fewer, though some will remain. The primary purpose of nuclear weapons will be political, to demonstrate to a potential aggressor the incalculable risk of initiating or continuing a war in Europe. What such a view implies is that for the very unlikely possibilities of both a conventional war and a prospective defeat, the West would need to hedge by maintaining the capability for a limited number of initial nuclear strikes against targets associated with the conventional battle. Those strikes could be planned to affect the immediate battlefield situation and/or to disrupt any reinforcements. For deterrence of a nuclear attack in Europe, the West would continue to depend on its ability to threaten a massive nuclear response with U.S. strategic nuclear weapons. The military role of the U.S. nuclear weapons stationed in Europe would be simply to ensure escalation through limited retaliatory strikes on the military forces and territory of the aggressor.

Having determined the tasks for nuclear weapons, the West will need to consider whether land-based nuclear systems will remain necessary. Past arguments for land basing have included the need for systems of every variety, to ensure that the Soviet Union would not believe itself safe from any kind of retaliatory attack. Land-based nuclear systems, such as short-range nuclear missiles and dual-capable aircraft, demonstrate visibly the U.S. commitment to the defense of Europe. Some believe an American president is more likely to use land-based systems because the damage to the United States might be minimized and attacks on the superpower homelands would be postponed. Some also argue that deterrence depends on the Soviet Union's understanding that its homeland is vulnerable to attack from other than U.S. strategic systems.

The contrasting view is that a credible deterrent does not depend on whether weapons are located on land or at sea. Rather, it is the U.S. commitment of its nuclear and conventional forces to Europe that deters. A decision to use nuclear weapons will be a function of how Americans assess their interests and vulnerabilities, not the particular weapons to be employed. The potential target and the American ability to attack it will be more important than where the weapon is based.

Additional political and military arguments have been made in the past for land basing. American officials have suggested that the United States would not be willing to extend its nuclear umbrella to Europe unless Europeans are prepared to share the risk by basing U.S. nuclear weapons on their own territory. The command and control of sea-based systems is less secure than those of land-based systems, and their communication much less timely. Firing a limited number of submarine-launched ballistic missiles (SLBMs) will give away the location of a submarine, jeopardizing its role as a secure strategic platform. Attack submarines with submarine-launched cruise missiles (SLCMs) that are carrying out their antisubmarine role could well be out of place to launch strikes into the European theater. SLCMs on platforms dedicated only to SACEUR would be very expensive.

Finally, there is the issue of what role British and French nuclear forces should play in preventing war in Europe. Both governments remain committed to retaining their nuclear forces as the ultimate guarantor of their own security. Their forces could also deter threats against other European states, either independently or with weapons of the United States.

Assuming that the West still sees a role for U.S. nuclear forces in Europe, three approaches for the West's nuclear strategy can be defined from these various views and considerations. One approach would be to maintain a broad spectrum of nuclear capabilities, so as to guarantee their survival under the worst case of a surprise attack, to blunt a conventional attack, and to provide for nuclear escalation. Land-based systems would be needed to ensure the credibility of the American nuclear guarantee and to overcome the limitations inherent in sea-based systems. They would also continue to involve Europeans in the sharing of nuclear risks. The United States would deploy a new tactical air-to-surface standoff missile (TASM). Reductions in the U.S. nuclear stockpile would be made over time, consistent with the reduction in the military threat, to some 2,000 warheads.

A second approach would be based on the judgment that modernized land-based nuclear systems remain necessary, but that a credible deterrent will require only a minimum deterrent force with the capability for a small number of initial and retaliatory strikes. The nuclear force posture for this strategy would include some 500–800 weapons, but they would need to be survivable and militarily effective. Because of political opposition to ground-based missiles and artillery, such a minimum deterrent would rely primarily on sea-based systems as well as a new TASM. British and French nuclear weapons could be part of this deterrent.

A third approach would be to place primary reliance on sea-based systems. SLBMs would provide for limited escalatory strikes, and SLCMs for attacks against targets in the theater, including some related to the conventional battlefield, such as airfields and command and control sites. U.S. land-based systems (ground and air weapons) would not be modernized. Under this approach the role of British and French nuclear aircraft and perhaps land-based missiles could be expanded for the defense of the other European states.

The choice of an approach involves fundamental judgments as to what is necessary to ensure a credible deterrent to war in Europe. Having agreed on a common approach, the West would be in a position to define both its modernization program and its objectives in the promised arms-control negotiations.

Future Directions in Nuclear Arms Control

Public pressures in the West led NATO in May 1989 to agree, in principle, to negotiations on short-range nuclear forces. NATO governments confirmed their commitment in the spring of 1990. With subsequent political developments in Europe, is there still a case for arms control? Or should the West proceed with unilateral reductions, calling upon the Soviet Union to reciprocate? Both superpowers have indicated a willingness to reduce their respective theater nuclear forces. Negotiations will take time and could hold such reductions hostage to an agreement. The verification provisions will be particularly difficult to design because most of the short-range systems are dual capable.

The process might be started with the United States and the Soviet Union each making a declaration, now that a CFE agreement has been signed, as to the nuclear weapons it plans to maintain in Europe with its reduced conventional forces, as well as the number of nuclear weapons to be reduced unilaterally. Through a bilateral Soviet–American working group, each side would provide the other with the means to ensure that the unilateral reductions were indeed being made. They could include, for example, observers to monitor the removal of weapons from storage sites.

If negotiations remain a priority, so as to ensure mutual and verifiable reductions, then it will be necessary for the West to define its objectives. Views differ as to whether arms control has a role to play in ensuring a credible nuclear strategy. There are those who argue that the West's nuclear requirements are a function of what is necessary to deter and are essentially independent of the size and character of the Soviet threat. The alternative view is that arms control, if it succeeds in reducing the Soviet threat, would affect NATO's own requirements as in the 1979 dual-track decision and in the INF treaty. NATO leaders in May 1989 did not explicitly offer a view, stating only that a minimum number of nuclear weapons must be maintained, but mutual reductions to equal and verifiable levels of short-range nuclear forces were desirable.

In strategic terms an agreement that called for the elimination of all short-range missiles would require significant reductions in the Soviet arsenal. Such an agreement would be attractive to those who have argued in the past that superiority in these and other theater nuclear weapons has seriously undermined NATO's escalation dominance, or the ability to win at every level of escalation. For those who worry that ballistic missiles are instruments of surprise attack, an agreement that reduced the approximately 1,600 Soviet shorter range ballistic missiles deployed in the Atlantic-to-the-Urals area would contribute measurably to stability.

If the objective is to reduce the threat and Soviet superiority, then the negotiations should instead cover all theater nuclear forces (ground and air missiles as well as artillery). Expanding the scope will also be attractive to Western public opinion, which favors reductions in all nuclear arms. The unit of limitation will need to be missiles and artillery shells, rather than launchers or aircraft, if the goal is to achieve real reductions in nuclear weapons or to avoid constraints on conventional capabilities. But limitations on nuclear missiles alone will be extremely difficult to verify, given that a missile with a nuclear warhead is virtually indistinguishable from one with a conventional warhead.

If the goal is to limit ground-based systems, a geographic region confined to the Atlantic to the Urals, as in CFE, would adequately address the threat to the United States and its allies in Europe and Asia; but if air-delivered weapons are included, then pressures can be expected from Japan for a global ceiling, as in the INF treaty.[3] The actual negotiating objectives must also be linked with the strategy and force posture that has been adopted for U.S. nuclear forces in Europe.

More fundamental problems will, however, arise in negotiating an agreement on short-range nuclear weapons beyond those involved in determining the negotiating objectives or the modernization and arms-control linkages. Great Britain and France also deploy short-range nuclear weapons, and both are adamant in their opposition to including these or their strategic forces in any future negotiations. Each has set stringent conditions for even considering the possibility, the most important being their insistence on substantial reductions in the strategic arsenals of the superpowers and no significant changes in Soviet defense capabilities. Even if these conditions were met, the two governments are still likely to resist including their national forces in East–West negotiations. The British and French each have nuclear weapons on their aircraft. The British deploy the *Lance* missile with U.S. warheads, and the French have their own short-range nuclear missiles. Whether the Soviet Union will finesse limits on British and French nuclear forces, as it did in the INF treaty, is most uncertain.

A more fundamental problem is that Western public opinion will be attracted to further zero options and the elimination of additional classes of nuclear weapons. The Soviet Union rhetorically seeks the elimination of all U.S. nuclear weapons in Europe and, in the INF negotiations, demonstrated its unwillingness to legitimize in an arms-control agreement any American deployment of nuclear weapons in Europe. While there are indications that Soviet leaders may be prepared to accept a minimum nuclear deterrent in Europe, they certainly cannot be expected to make the case for nuclear weapons in the Western public debate.

CONCLUSIONS

The Future of NATO

As the Cold War ends, the West is poised to realize its postwar political goals. NATO as a military alliance was founded to contain the spread of communism and to prevent the expansion of Soviet political influence and military power into Western Europe and has succeeded over forty years in achieving these goals. Now the threats themselves are disappearing. At the same time, the ideological and political cohesion of the Warsaw Pact has been shattered, and its military role ended. Gorbachev has retreated from his call for a dissolution of blocs, claiming that the Warsaw Pact can help promote stability in Eastern Europe. Fears of Germany may lead others in Eastern Europe to the same view. NATO, too, is now being portrayed as a means of ensuring stability, by anchoring Germany in the West and the United States in Europe.

Institutions such as NATO and the Warsaw Pact cannot, however, promote stability in the abstract. Withering alliances will not—indeed, should not—prevent radical political change in Europe. NATO needs to be maintained for as long as it serves legitimate military purposes and should be defended publicly in this way. In the coming years NATO will be involved in structuring military forces under arms-control agreements and in monitoring their implementation. It could play a role in managing crises if political turmoil in the East were to threaten the peace. It is a hedge against a resurgent Soviet military threat. But NATO should not search desperately for new roles if its military purposes disappear. It should not try to compete with other institutions, such as the EEC and the CSCE, which are clearly more appropriate for promoting the West's political and economic goals in Europe. NATO is useful as a forum for political consultation, but certainly not unique, and consultations are not an end in themselves. Americans may wax nostalgic over NATO, as it has provided them with a commanding leadership role, but developments in Europe are changing that role as well.

The critical issue for the future of NATO is the future of Germany. While compatible with Germany remaining in the EEC, reunification could well over time be incompatible with membership in NATO because of objections from the Germans themselves. The West must plan for the possibility that it will not be able to achieve its twin goals and must bear in mind in its efforts to find a way to keep Germany in NATO the longer term risks of isolating the Soviet Union within a future European security system. If Germany eventually leaves NATO, would the alliance retain a legitimate military or political purpose? There will still be the need to balance Soviet military power in Europe. The problem is that NATO under these circumstances will look like an alliance directed more at Germany than at the Soviet Union. The future security structure in Europe needs to include both the Soviet Union and Germany, even if it means the end of NATO.

The Strategic Purposes of Conventional Forces

Political pressures in the West are understandably rising for significant reductions in conventional military forces, particularly in the United States and Germany. The strategic debate has unfortunately become preoccupied with rather abstract concepts of defensive restructuring and strategic mobility. The idea of multinational forces is again in vogue, this time with units deployed not only in Germany but throughout Western Europe. What is missing is a clear sense of why the West collectively, or individual nations, will wish to maintain conventional forces in Europe, given that the Soviet military threat has all but disappeared.

Among the potential strategic purposes, a military presence could promote stability given the possibility that ethnic and nationalistic conflicts may arise in the East; could provide a hedge against the reconstitution of a Soviet military threat; could extend security guarantees to the democratic states of Eastern Europe; and could ensure through a minimum level of military forces the independence and integrity of the individual nation-states. The military requirements for each of these would be very different, as would be their importance to each nation.

The prospect of conflicts in Europe has not disappeared with the end of the Cold War. Violence is a real possibility in the East, where nationalistic and ethnic rivalries abound. Apart from ending the bloodshed, Western governments will be interested in ensuring that a conflict does not spill over and involve the major European powers. The difficulty is in finding any role for Western military forces. Peacekeeping operations are a possibility, but their success would depend on whether those involved in the conflict came to prefer peaceful means to violence in promoting their interests, and whether they would be willing to accept mediation and military intervention on the part of any Western countries. The number of military forces for such peacekeeping operations would be relatively small, with a potential military response tailored to each crisis as it arose.

A reconstituted and serious Soviet military threat looms as a long-term possibility. The military forces required so that the West could respond might be very large, but there would be sufficient time to prepare a successful conventional defense. Moreover, with the withdrawal of Soviet military forces from Eastern Europe, the Soviet Union would face a formidable task in mounting a major attack even against Germany. There is always the possibility of Soviet aggression along its border on NATO's flanks, and the Soviet Union might seek to reinstitute control in Eastern Europe. In the first case, the West would be called upon to respond militarily; in the second case, it would be more difficult than in the past to stand aside, although the West would wish to avoid any explicit security guarantees of the countries of Eastern Europe. In both cases, the Soviet Union would have to expect serious resistance within the countries themselves, and it is difficult to posit what political objectives they would see being served.

So Western governments face the task of defining the minimum levels of military forces to ensure their own national independence while hedging against future uncertainties and instabilities as well as the long-term possibility that a serious military threat could reemerge.

The difficulty of this task will lead almost inevitably back to arms control, where the objectives seem much easier to define: reductions in military forces to equal levels (as will be accomplished by implementation of the CFE treaty) and perhaps over time a defensive restructuring of the military forces of the various countries. Negotiations also buy time. But the problems with arms control are obvious. The outcome is left to what is negotiable and, therefore, acceptable to one's potential adversary. It will not necessarily provide for any of the potential strategic purposes required of a conventional military presence in Europe. Negotiations hold hostage reductions in military forces when reductions no longer seem to be needed. Most importantly, with the political developments in Europe, the basis for setting equal ceilings or defining limits on the military forces of blocs or even adversaries has essentially disappeared. Suggestions simply for percentage reductions from the CFE levels in the military forces of individual nations might be politically attractive, but they would at best be cosmetic and could not be expected to respond to pressures in either Germany or the United States for significant reductions.

Arms control, therefore, cannot and really should not be the means of establishing future levels of conventional forces in Europe or, more indirectly, their strategic purposes. Nations will need to do this individually, judging what, if any, threats exist to their security in light of developments in the Soviet Union and Eastern Europe. The diminished threat provides an opportunity for significant reductions in military forces. Governments will also wish to adopt a more defensive military strategy, placing greater reliance on mobility and reserves. Having determined unilaterally the levels of military forces, countries should then announce what these would be and agree to their being confirmed and monitored under the verification measures to be put in place in the CFE agreement. But these levels would in no way be mutual, collective, or necessarily equal to those in any other country. The national plans could be discussed and even coordinated within NATO, if that were the wish of the alliance members. Multinational forces could also be created within NATO or among various European countries.

The major argument against such unilateral restructuring will arise in the context of the future level of military forces to be maintained in Germany. Chancellor Kohl, in response to Soviet pressures, has agreed to a ceiling on the Bundeswehr of 370,000 ground and naval personnel. The follow-on negotiations in CFE are now viewed by the Germans as a means of avoiding being singled out for special arrangements. The proposal now in vogue is to extend the sufficiency rule beyond armaments so that no single country could have more than a certain percentage of the overall level of manpower permitted each bloc. The problem is that such a formula would still single out Germany. And there is now no Warsaw Pact bloc to which to assign reciprocal ceilings.

The best approach would be for the new government of Germany to an-

Beyond German Unification

nounce voluntarily its proposed ceiling on the Bundeswehr and to agree to having that residual level confirmed and monitored under the verification regime to be put in place under CFE. All other Western governments would announce their readiness to make similar declarations as to their future levels of conventional forces and to accept similar monitoring. The Germans would not be singled out for any special arrangements, and they would have responsibility for defining their own future force posture. Such a step would, in effect, meet Soviet concerns without sowing the seeds of resentment within Germany. Negotiated arms control is not the best way to proceed with respect to establishing the future size of the military forces in Germany, or for any other countries in Europe.

The Role of American Troops

The primary arguments for withdrawing American forces from Europe are political—to respond to domestic pressures in the United States to reduce the defense budget and to shed the burden of providing for Europe's security when the military threat has essentially disappeared and Western Europeans can afford to pay for their own defense. The Soviet Union may still press for mutual Soviet–American reductions so as to make their own withdrawal from Europe politically more acceptable. Over time, Europeans, and especially the Germans, may not be willing to accept the stationing of American forces. The question is whether these various political pressures should be resisted.

From a coldly analytical point of view, there is no reason to keep American conventional and nuclear forces in Europe. The United States is intimately tied to Europe, culturally, politically, and economically; military forces are not needed to ensure that bond. Europeans in peacetime could balance the military power of a weakened Soviet Union. And the United States would be able to mobilize its military forces and return to Europe quickly in a crisis.

But symbolically, American troops demonstrate American interests in Europe, and again all the lessons of history suggest that some should remain, not only for the future stability of Europe, but for the protection of America's own security. Americans twice in this century returned to Europe with military forces too late to keep the peace.

The role of American military forces would be primarily to balance Soviet power in Europe, but at the same time they would provide a deterrent to any state seeking to expand its power directly through military force or indirectly through political coercion and intimidation.

Again, this role could be played from the United States, but it raises as potential problems those that England confronted when seeking to balance Germany and deter its expansion earlier in this century. The British government had difficulty finding friends and forging alliances quickly enough, as with France before World War I. It failed to credibly demonstrate its interests by having its military forces in place on the Continent when Germany invaded Czechoslovakia and Poland before World War II. A minimum military presence in Europe, defined through bilateral or multilateral security treaties, would ensure that American power and interests were viewed by all the states of Europe as involved in their future and that an American military presence was there to deter specific threats to peace from wherever they might arise.

The presence of some American troops is what is important, not the numbers or whether they are ground or air forces. Equating Soviet and American forces in CFE was a mistake, for it has made it more difficult to call for the removal of all Soviet military forces from Eastern Europe while maintaining some American troops in Western Europe. The actual number of American forces in Europe will be arbitrary, by definition. But the remaining forces should be so structured as to be militarily credible and not simply to provide a logistics and communication base.

Conventional Arms Control

The CFE agreement has been completed. As for the future of arms control in Europe, negotiated agreements should not be abandoned, as the possibility of war in Europe will not disappear. But agreements need to be designed to meet the potential threats to peace and stability, and their focus will need to be on CSBMs within the CSCE framework. America has in the past opposed any CSBMs which would constrain military purposes. In principle, however, CSBMs appear attractive as a means of providing for the defensive restructuring of the residual military forces of the various countries and of covering naval activities as well as the global military activities of the two superpowers.

CSBMs might also provide the basis from which to design a mechanism under the CSCE for crisis management. If a serious crisis were to arise in the future, it will almost certainly involve changes in the disposition of military forces. These changes could trigger an inspection under the CSBM regime and could perhaps even constitute a violation of one of the measures. These changes would certainly provide a reason for political consultation, and could prompt collective diplomatic and military action to defuse a crisis. In this way, CSCE and CSBMs could evolve naturally as a framework for managing crises in the future.

U.S. Nuclear Weapons in Europe

Political pressures now seem almost destined to lead to the removal of most, if not all, U.S. nuclear weapons from Germany—and perhaps from all of Europe. The Soviet Union may insist upon limits on nuclear weapons (and perhaps their elimination) in Germany. Antinuclear feelings remain widespread in Europe. The strategic arguments of the past for maintaining U.S. nuclear weapons in Europe are disappearing.

Rather than addressing the strategic issue of whether there remains a role for U.S. nuclear weapons, Western governments continue to focus on individual nuclear-weapons systems. President Bush had an opportunity in the spring of 1990 to offer a strategic view but instead simply cancelled two weapons, neither of which the Germans were willing to deploy on their territory.

NATO returned in July 1990 to the same language of past communiques: "The Alliance must maintain for the foreseeable future an appropriate mix of nuclear and conventional forces based in Europe, and kept up to date where necessary." It then went on to conclude that there will now be "a significantly reduced role for sub-strategic nuclear systems of the shortest range," and in the transformed Europe nuclear weapons will be "truly weapons of last resort." NATO has also proposed "in return for reciprocal action by the Soviet Union, the elimination of all its nuclear artillery shells from Europe."[4]

The fundamental strategic question is whether U.S. nuclear weapons should be maintained in the future for the residual purpose of promoting stability and preventing war in Europe. Nuclear weapons do not contribute to stability in the abstract. They exist in a political setting and must be related to particular military threats. But by making the potential risks of a war in Europe unacceptably high, they do introduce caution into the actions of government leaders. That caution is important so as to deter the use of military force not only by the Soviet Union but also by all other states in Europe. However incredible the threat to use nuclear weapons may seem, their existence necessarily affects potential decisions on using military force. It is the fact of nuclear weapons that makes the developments in Eastern Europe today, and the possibility of potential ethnic and nationalistic conflict, different from those of the 1920s and 1930s.

There is a critical residual role for U.S. nuclear weapons in the future to promote stability and prevent war in Europe. American nuclear weapons can also provide an ultimate security guarantee for Germany, assuming that it remains linked politically and economically with the West and continues to forgo any nuclear weapons of its own.

The West needs, therefore, to adopt a strategy of minimum deterrence. Such a strategy would involve maintaining a small number of U.S. nuclear weapons in Europe, a stockpile of perhaps 500, that would serve the primarily political purpose of influencing the calculations of costs and gains in the potential use of military force. This has historically been how Europeans have viewed the role of nuclear weapons and would approach the French existential view. But Americans will continue to insist upon flexibility and militarily effective options. So the strategy of minimum deterrence must still be one of flexible response.

In the promised negotiations, the United States would seek reductions in the Soviet nuclear arsenal in Europe, placing a ceiling on land- and air-based nuclear weapons from the Atlantic to the Urals at an equal level consistent with this minimum deterrent. The ceiling would be on nuclear weapons, thereby giving each side the freedom to define the characteristics of its own force posture, providing flexibility for modernization and changes over time. Categories of weapons would not be limited, and there would be no further zero options.

The West would thereby test Soviet rhetoric in favor of significant reductions in its theater nuclear arsenal and find out whether the Soviet Union is prepared in an agreement to legitimize some minimum number of American nuclear weapons in Europe. If so, the West will have gained, at least implicitly, Soviet acceptance of its view that nuclear weapons contribute to peace and stability, which should help to gain the support of Western public opinion for maintaining a minimum nuclear deterrent. If, as in the INF negotiations, the Soviet Union refuses to agree to any outcome other than zero, it would risk losing the support of Western governments for its broader political and economic goals.

Clearly, a minimum deterrent must be based on modern and survivable nuclear weapons. It would include air- and sea-based systems, with the mix to be determined by what is politically feasible and militarily most effective. Arguments made in the past for land-based as opposed to sea-based systems, as symbols of American and European commitments, no longer seem appropriate when nuclear weapons serve as simply the ultimate guarantors of peace. As the number of nuclear weapons will be based on a judgment as to what is necessary for deterrence, and not on the size of the Soviet theater nuclear arsenal, reductions in older artillery and land-based missile systems should take place irrespective of what happens in the upcoming negotiations—and should not be held hostage to their success. Given that the Soviet Union ostensibly favors the elimination of all nuclear weapons, these older American nuclear weapons will not be required as bargaining chips.

Where the nuclear systems will be based and what their relationship will be to NATO will depend on how Europe evolves politically. Ideally, some few nuclear weapons would remain in each of the countries where they are currently based. British and French nuclear weapons would contribute to this minimum deterrent, sharing responsibility with the U.S. weapons for preventing war in Europe. Planning for these nuclear weapons would be for limited initial and retaliatory strikes against the military forces of a potential aggressor, whoever that might be.

Reconstructing the Foundations of Peace

Rarely has history produced so revolutionary, yet so peaceful, a change as has occurred in Eastern Europe and the Soviet Union. Breathtaking in their pace, the events are sobering in the uncertainties they raise. The task of defining the

West's military strategies and arms-control policies has become even more difficult, but certainly no less important. The challenge to the West is to seize the opportunities presented by developments in the East without losing the stability and peace of the past.

What this means is maintaining a viable NATO alliance as long as it has legitimate military purposes, but not permitting it to stand in the way of ending the division of Europe and of promoting East–West economic and political cooperation through other more appropriate institutions. Governments should proceed with unilateral reductions and restructuring of their military forces consistent with their view of the threat to their own security and how it evolves. But the process of arms control should not be abandoned.

Most important, the START agreement needs to be concluded and implemented, and the CFE treaty implemented, so as to establish parity in strategic nuclear and conventional forces in Europe. The United States should take the initiative now to ensure further significant reductions in Soviet nuclear forces, in START and in negotiations on short-range nuclear forces. Arms control also has a role to play in Europe in promoting stability as the political and military structures evolve, but when the threats to peace have not disappeared. CSCE is most attractive because it focuses on CSBMs and links political and economic goals with security for the whole of Europe.

How Europe evolves will ultimately depend on the aspirations and interests of the individual countries. There is no proof that maintaining a minimum nuclear deterrent and the presence of American troops will determine whether the outcome is cooperation rather than conflict. But both are prudent hedges, and history suggests it would be wise to preserve them so as to ensure peace in a changing world.

NOTES AND REFERENCES

1. This chapter draws extensively upon the analysis and recommendations in the author's recent study, *Assuring Peace in a Changing World: Critical Choices for the West's Strategic and Arms Control Policies* (Washington, D.C.: The Johns Hopkins Foreign Policy Institute, 1990).
2. Klaus Wittmann, "Challenges of Conventional Arms Control," *Adelphi Papers* 239 (London: International Institute for Strategic Studies, Summer 1989).
3. This discussion of the objectives for negotiations on short-range nuclear forces draws upon William D. Bajusz and Lisa D. Shaw, "The Forthcoming SNF Negotiations," *Survival* (July–August 1990), pp. 333–47.
4. *The Washington Post*, July 7, 1990, p. A18.

8

American Security Policy in the Pacific Rim[1]

Harry Harding

The present strategic environment in the Pacific Rim consists of four interconnected elements, which have emerged at various times since the end of World War II. These four elements, which will continue to shape the East Asian scene throughout the 1990s, include:

- The regional disputes in Korea, the Taiwan Straits, and Indochina, in which the United States has had an interest since the late 1940s and early 1950s, but in which the American role is gradually becoming less central and less direct.
- The rise of the Soviet Union as a significant military power in East Asia in the mid 1960s, followed first by the creation of a loose anti-Soviet united front in the 1970s and then by the moderation of Soviet policy in the region under Mikhail Gorbachev in the mid 1980s.
- The economic dynamism that spread across most of the region in the 1970s and 1980s, producing greater interdependence and competition and raising economic issues to higher prominence on both domestic and international agendas.
- The gradual emergence, beginning in the 1980s and continuing into the next century, of a number of significant independent regional powers, including China, India, and Indonesia, and the resulting pressures toward a more multipolar balance of power in the Asia–Pacific region.

The first section of this essay offers an overview of these four elements in the East Asia strategic environment and their implications for American interests.

Harry Harding • The Brookings Institution, Washington, D.C. 20036.

The second section considers the challenges and opportunities that the four trends pose for various aspects of U.S. security policy in the region, including American military strategy, force posture, arms-control policy, diplomatic strategy, and alliance management.

My principal thesis is that in the 1990s the United States will require a more comprehensive view of the strategic situation in the Pacific Rim. In the past the United States was understandably preoccupied with the first two elements in its Asian agenda: its involvement in the regional disputes in Korea, Indochina, and the Taiwan Straits and its efforts to counterbalance the expansion of Soviet military power in the Asia–Pacific region. Our interests in these two areas remain important and will demand our continued attention. But the United States must now supplement these traditional concerns with a growing attention to the latter two aspects of the Asian security environment: the economic dimensions of national security and the emergence of a greater diffusion of power along the Pacific Rim. Unless we do so, our policy in East Asia will become increasingly obsolete—unable to maintain the support of our allies and unable to cope with the challenges and opportunities confronting the United States in one of the world's most dynamic regions.

THE ELEMENTS OF THE STRATEGIC SITUATION IN EAST ASIA

We turn first to a fuller consideration of the four key elements of the Asian security environment and their broad implications for American interests in the region.

The Evolution of the Three Regional Disputes

Ever since the early 1950s the United States has had an interest in three regional disputes in East Asia: the conflict on the Korean peninsula, the competition between Taipei and Beijing in the Taiwan Straits, and the ongoing effort by Hanoi to establish its preeminence first in Vietnam and then throughout Indochina. In each case the underlying dispute was the result of the division of the country after World War II, through either occupation or civil war, into Communist-controlled areas and areas controlled by non-Communist governments. The U.S. involvement in these regional conflicts stemmed from an American decision to protect the non-Communist part of the country against attack by its Communist neighbor.

Until the mid 1970s the American stake in each of these disputes was immediate and direct. In all three cases the United States had a formal security obligation to its non-Communist ally: a mutual defense treaty with South Korea, a similar treaty with Taiwan, and a commitment to South Vietnam through the Southeast Asia Treaty Organization. In each case, too, the United States had proven willing to use armed force in pursuit of its interests: in the Korean War of 1950–1953, in the Taiwan Straits crises of 1954–1955 and 1958, and in the Vietnam conflict from 1964 onward. As of 1974, American forces were stationed in all three conflict areas.

Moreover, the United States also dominated the diplomatic channels designed to reduce tensions or resolve these three disputes. In the absence of any formal contacts between Beijing and Taipei, it was the United States that talked with mainland China about the situation in the Taiwan Straits. Although both Seoul and Washington had representatives at the armistice negotiations at Panmunjom, and subsequently at the armistice commission, it was the United States that dominated the dialogue. Similarly, although Saigon participated in the Paris Peace Conference on Vietnam, again it was Washington that set the agenda.

Over the past sixteen years, however, the nature of the American military involvement in these three disputes has become less central. As part of our agreement on establishing formal diplomatic relations with China, we completely withdrew our forces from Taiwan in 1979 and terminated our mutual defense treaty with Taipei at the beginning of the following year. Ever since the collapse of South Vietnam in 1975, the United States has had a very minor involvement in Indochina, consisting of small amounts of financial aid, material assistance, and diplomatic support for the non-Communist resistance forces in Cambodia. Although we maintain our commitments to South Korea, the number of American troops stationed in that country is significantly less now than in the late 1960s and is to be reduced further by approximately 10 percent.

The American involvement in the processes of tension reduction and conflict resolution in these three regional disputes has also undergone a significant change. As our military involvement has become less direct, we have simultaneously begun to emphasize the importance of dialogue and negotiation between the parties directly involved. In keeping with such an approach, we have supported the resumption of a political dialogue between North and South Korea; promoted international negotiations on the future of Cambodia; endorsed the Association of Southeast Asian Nations (ASEAN) position in its negotiations with Vietnam; and welcomed the growing economic, academic, and humanitarian ties across the Taiwan Straits. Washington has increasingly acknowledged that the United States can no longer dominate these contacts and negotiations. Instead, the solutions to Asian regional problems can only be those that Asians themselves have worked to develop, with outside powers playing the role of catalyst and guarantor.

Fortunately, the new American emphasis on seeing solutions emerge from the inside out, rather than attempting to impose them from the outside in, has been facilitated by the growing flexibility of most of the parties to these conflicts. Vietnam has withdrawn most of its military forces from Cambodia, and there is growing consensus on the desirability of a four-party interim government in

Phnom Penh to conduct national elections under United Nations supervision. Beijing has offered to negotiate with Taipei over the terms for China's reunification and has indicated that it would use force against Taiwan only in the most extreme circumstances. Although still reluctant to engage in direct negotiations with Beijing, Taipei is willing to tolerate—even encourage—the rapid expansion of trade, investment, cultural and academic exchange, and family visits across the Taiwan Straits. Political dialogue on the Korean peninsula has now resumed, and there has been progress toward expanding economic and cultural relations between the two Koreas. In each case the increasingly competitive economic environment of the Asia–Pacific region appears to be encouraging greater flexibility, as the parties to the disputes seek to reduce the costs of confrontation and gain the benefits of economic cooperation.

There are, of course, serious uncertainties in each of these three regional conflicts. North Korea still seems intent upon enhancing the size and readiness of its armed forces, with some recent reports even suggesting that Pyongyang may be attempting to develop nuclear weapons. If reform in China should falter, if Hong Kong's return to Chinese sovereignty should prove chaotic, or if pressures for independence should build in Taipei—all possible consequences of the repression of popular protest in China in June 1989—then the chances for a peaceful resolution of the Taiwan question would be substantially reduced. Despite the progress in the negotiations on Indochina noted above, there is as yet no agreement on the structure of an interim government in Cambodia, the place of the Khmer Rouge in such a regime, or the role of the United Nations in guaranteeing the implementation of an agreement. Nor is there any certainty that renewed civil war in Cambodia can be prevented after elections are conducted.

Still, the prospects for each of these three regional disputes appear better today than at any time in the last three decades. At a minimum the tendencies are toward a gradual reduction of tensions and the avoidance of military escalation. In some instances there is now the possibility of building economic and cultural contacts across what once was an unbridgeable political divide. At least in the case of Cambodia, there now exists the chance of progress toward a comprehensive negotiated solution.

The Rise and Retreat of the Soviet Union

The second element in the present strategic environment of the Pacific Rim, the emergence of the Soviet Union as a major military power in East Asia, has occurred in several stages over the last forty years. Until the early 1960s, Soviet power in the Asia–Pacific region was exercised not so much by its own military deployments, but through the Kremlin's cultivation of various proxies and surrogates in the area. China was, of course, the Soviet Union's principal ally in the region in the 1950s. But close links with other Communist countries, including

Mongolia, North Korea, and North Vietnam, also helped extend Soviet influence, as did the Communist parties operating through much of South and Southeast Asia.

The Sino-Soviet conflict, which broke out in the late 1950s and escalated rapidly in the early 1960s, removed the prospect that China would serve as an instrument of Soviet policy in East Asia. Moreover, as the exchange of ideological polemics evolved into a direct military confrontation along the two countries' disputed frontier, it also stimulated a significant increase in the size and sophistication of Soviet forces in the Far East, as well as the introduction of a contingent of Soviet ground and air forces into Mongolia. The clashes between Chinese and Soviet troops along the Amur and Ussuri rivers in 1969, and the rumors of a massive Soviet attack against Chinese military installations that summer, marked the full emergence of the Soviet Union as a major military power in East Asia.

The 1970s and early 1980s saw a further step in the development of Soviet strategy and deployments in East Asia. Moscow supplemented its direct defense of the Sino-Soviet border with efforts to encircle China both diplomatically and militarily, including the development of a strategic relationship with India, the inauguration of a Soviet alliance with Vietnam, the intervention in Afghanistan, the utilization of the Vietnamese naval base at Cam Ranh Bay, and the expansion of Soviet political and military ties with North Korea. During the same period, the Soviet Union also began to use the Far East as a base for portions of its strategic deterrent, with the installation of SS-20 ballistic missiles in Siberia, the stationing of nuclear-powered ballistic-missile submarines (SSBNs) in the Sea of Okhotsk, and the deployment of the forces required to defend those assets against a possible conventional attack by the United States.

The most controversial element of Soviet strategy in the Far East has been the development of projection capabilities in the region in the past decade, including aircraft carriers, *Backfire* bombers, and Marine contingents. The purposes to which those forces might be put have been the subject of considerable debate among Western analysts. Some have suggested that they might be used to cut the sea lines of communication in the Western Pacific, particularly those leading to Japan from the Straits of Malacca and the United States. Others have warned that they might be employed to seize and hold parts of northern Japan to serve as a bargaining chip in the event of a global conventional war with the United States. Still others have noted that Soviet forces could be used to attack American installations in the Western Pacific, as well as elements of the U.S. Seventh Fleet. But some analysts have insisted that the most plausible mission for Soviet conventional forces was simply the defense of Soviet military assets against preemptive attack by the United States.

Whatever its purposes, it became increasingly clear by the late 1980s that the massive buildup of Soviet forces was both expensive and counterproductive.

The cost could be calculated not only in monetary terms, but also in the numerous deaths and injuries suffered by Soviet forces in Afghanistan. Moreover, this sacrifice of financial and human resources did little to improve the Soviet position in East Asia. Instead, the expansion of Soviet forces created widespread apprehension about Moscow's intentions throughout the region. In the late 1970s and early 1980s, it even threatened to create a loose alliance of the United States, Japan, China, and the non-Communist states of Southeast Asia to resist Soviet hegemony. Although it maintained close ties with India and Vietnam and succeeded in improving its relations with North Korea, on balance the Soviet Union was isolated in East Asia as never before.

Faced with a situation in which a high price was being paid for a policy with poor results, Mikhail Gorbachev has attempted since 1985 to redefine Soviet strategy in East Asia in less costly and more productive ways. The cornerstone of Moscow's new Asia policy has been its attempts to improve its relations with China, primarily by making unilateral concessions on all three of the issues that were of particular concern to Beijing: Soviet involvement in Afghanistan, Soviet support for Vietnam's intervention in Cambodia, and Soviet force levels in Mongolia and Siberia.

Most Soviet analysts understand that there is little possibility of re-creating the Sino-Soviet alliance of the 1950s. But they believe it is realistic to avoid the rigid confrontation that characterized their relationship with China during the 1960s and 1970s. At a minimum the goal of Soviet policy has been to ensure that China would not join the United States and Japan in any anti-Soviet coalition. And as a maximum objective, Gorbachev has clearly hoped that Beijing might be willing to endorse Soviet arms-control initiatives, to support Soviet proposals for the resolution of various regional conflicts, or to reestablish military-to-military contacts between the two countries. By mid 1990 the Soviet Union had achieved considerable success along many of these dimensions.

Although China has been the main focus of the new Soviet policy in East Asia, Moscow has also made significant overtures in other parts of the region. The Soviet Union has attempted, in large part by greater flexibility on the Cambodian question, to improve its ties with the non-Communist nations of Southeast Asia. It has developed economic and cultural relationships with Taiwan and has established formal diplomatic relations with South Korea; it has encouraged some of its allies, notably Mongolia and Laos, to be more flexible in their policies toward Japan and the West; and it has pressed both Vietnam and North Korea to adopt more conciliatory approaches to the resolution of the regional disputes in which they are directly involved. Moscow would also like to improve its relations with Japan, and has begun to hint at concessions on the territorial issues that Tokyo has identified as the most important obstacle to warmer Soviet–Japanese ties.

As in other parts of the world, Gorbachev's initiatives in East Asia are difficult to assess, because they are designed to advance two sets of objectives simultaneously. On the one hand, they are intended to reduce tensions with major actors in the Pacific Basin, including the United States, so as to ease the burdens of Soviet foreign policy. On the other, they are also designed to maximize Soviet leverage in its ongoing competition with the United States. Soviet policy in Asia, in short, can best be described as one of competitive accommodation with the United States and its allies in Asia, rather than as one of either confrontation or accommodation alone.

Consider first the more accommodative aspects of recent Soviet policy in the region. As already noted the Soviet Union has taken a more forthcoming and flexible approach toward virtually all the outstanding local disputes in East Asia. It has withdrawn its forces from Afghanistan. It has encouraged Vietnam to remove its forces from Cambodia and has accepted a four-power interim coalition government with an effective international peacekeeping force. The Soviet Union is also promoting a reduction of tensions on the Korean peninsula, both by expanding its own ties with Seoul and by encouraging various forms of dialogue across the demilitarized zone.

As an integral part of its program of economic *perestroika*, the Soviet Union has also expressed an interest in building cooperative economic relationships all across Asia. It is expanding its economic ties with virtually all Asian–Pacific countries. It has joined, or has stated its desire to join, a variety of regional economic institutions, from the Pacific Economic Cooperation Conference (PECC) to the Asian Development Bank. It is considering the establishment of special economic zones in the Far East to serve as a base for direct foreign investment from the nations of the Pacific Rim. Together with some analysts in China and Korea, Soviet observers have begun to discuss the possibilities of multilateral cooperation in Northeast Asia, which could link Japanese capital, Korean technology, Chinese labor, and Soviet resources. To be sure, the Soviet Union approaches East Asia with relatively few economic assets. Moscow can, however, offer some prospect of the export of natural resources and raw materials from Siberia to the rest of the region and of the import of at least limited quantities of capital equipment and consumer goods.

A further accommodative element of Soviet Asian policy has been the series of arms-control initiatives presented by Gorbachev over the past several years, first at Vladivostok in 1986, then at Krasnoyarsk in 1988, and finally during his visit to Beijing in 1989.[2] The more positive aspects of these proposals have been the inclusion of the SS-20s stationed in the Far East in a global ban on intermediate-range nuclear forces, a unilateral freeze on other Soviet nuclear forces deployed in Asia, the withdrawal of the bulk of Soviet forces from Mongolia, and the announcement of a reduction of Soviet troop strength in the Far East by some 120,000.

These positive developments, however, are but one side of the coin. It remains to examine those features of current Soviet Asian policy that pose a challenge to American interests. It is clear, for example, that the Soviet Union wishes to retain a significant strategic posture in the region. Moscow hopes to maintain strong political and military relationships with India and Vietnam. It still supplies advanced military equipment to North Korea. It would like, as mentioned above, to expand its military and strategic ties with China. It has continued to improve the equipment of Soviet forces in the Far East, even as it has announced plans to reduce their number. Finally, the dismantling of Soviet SS-20s in Asia is being counterbalanced both by the stationing of mobile SS-25s, which have the capacity to strike Asian targets, and by the redeployment of some Soviet SSBNs from off the coast of the United States back into Asian waters.

Moreover, some of the most dramatic of Gorbachev's regional arms-control initiatives include one-sided proposals that would affect American capabilities more sharply than Soviet resources. The most notorious instance is the Soviet proposal to withdraw from Cam Ranh Bay in return for American redeployment from Clark Air Force Base and Subic Bay Naval Station in the Philippines—a proposal that would, in effect, swap a relatively small Soviet installation for the cornerstones of American naval and air deployments in the Southwest Pacific. Other examples include long-standing Soviet support for nuclear-free zones in the Indian Ocean, Southeast Asia, and Korea, as well as Moscow's desire for agreement on naval arms control in the western Pacific.

In short, the more accommodative features of Soviet policy in East Asia provide some promise for a significant reduction of Soviet–American tensions in the region and for progress on the resolution of some outstanding local disputes. But other elements suggest that Moscow, committed to a competition for influence with the United States, is putting forward proposals that are designed to split Washington from its allies or to disproportionately reduce American military assets in the region. The challenge for the United States is how to respond to this complex Soviet strategy in ways that will secure its benefits at an acceptable cost.

The Economic Dynamism of the Region

The third major element in the Pacific Rim strategic environment has been its extraordinary economic dynamism, which has not only reshaped the domestic societies of virtually every country in the region, but has also begun to recast the relationships among them. The economic miracle of East Asia occurred first in Japan in the 1950s and 1960s and then spread to South Korea, Taiwan, Hong Kong, and Singapore in the 1970s. The economic successes of Northeast Asia are now beginning to be echoed in other parts of the region, including particularly Malaysia, Thailand, and parts of China. Given the proper domestic political climate, there is no reason why the miracle cannot be repeated in Indonesia and the Philippines as well.

The fact that East Asia has enjoyed much more rapid economic growth than any other region of the world has had a complex impact on international economic relationships there. On the one hand, there is a growing interdependence in the area, with a rapid expansion of intraregional trade, a steady increase in intraregional investment, and the emergence of various regional forums for consultation and dialogue on economic issues. On the other, economic dynamism is also creating a new sense of mutual competition throughout the region. Given the more rapid diffusion of technology and the expanding abilities of many Asian economies to absorb more advanced technology, countries that once were confident of their superior positions are feeling more intense competition from other countries that they previously thought were well behind them. Thus, Japan faces competition from South Korea and Taiwan in capital goods and electronics, South Korea and Taiwan perceive a growing threat from Southeast Asia in consumer goods and textiles, and the Southeast Asian nations encounter stiff competition from China in exporting manufactured goods to the West.

What are the ramifications of these developments for the national security interests of the United States? Increasingly, American analysts are exploring the consequences for U.S. security of the scientific, technological, and financial capabilities of Japan. Can the United States maintain an adequate manufacturing base for its defense industries in the face of strong Japanese competition? What are the security implications of the growing American dependence on imports of financial capital from a number of Asian countries? What are the consequences of possible Japanese dominance of key technologies, including new materials, supercomputers, and microelectronics? As important as these questions are, they are beyond the scope of this essay. Instead, we will focus here on the broader geopolitical implications of economic dynamism for the Asia–Pacific region as a whole.

To begin with, there is throughout the Pacific Rim a growing preoccupation with economics. The growing interdependence and competition mentioned above have meant that the issues of sustaining vigorous growth, identifying comparative economic advantage, increasing national economic productivity, and defining national economic strategy are now at the top of the domestic political agendas of virtually every country in the region. Relatedly, economic issues are now much more prominent in almost every bilateral relationship than was true ten or fifteen years ago. Then, for example, strategic and diplomatic questions dominated American relations with Japan, South Korea, Taiwan, China, and Southeast Asia. Today the most pressing issues center around the American desire to improve its current account balances with its Asian trading partners.

The growing prominence of economic issues has begun to complicate American relations with almost all of its friends and allies in the Asia–Pacific region. Moreover the scope of those issues is steadily expanding. Fifteen years ago the main problem was how to restrict the growth of Asian exports of textiles, shoes, consumer electronics, and automobiles into the United States. Now the economic issues between the United States and its Asian trading partners include the even more controversial issues of enlarging markets for American agricultural products, improving market access for American service industries, enhancing the climate for American investment in Asia, and monitoring (and possibly regulating) Asian investment in the United States. At the same time, controversies between the United States and various Asian nations no longer simply involve the issues of tariff and nontariff barriers for sensitive commodities. Today the questions in dispute are just as likely to involve exchange rates, interest rates, government budget deficits, savings rates, wholesale practices, retail networks, agricultural subsidies, and other issues traditionally regarded as matters of domestic economic policy and structure.

Great efforts have been made to prevent these economic disputes from affecting security relationships, but it is increasingly evident that America's military ties cannot be completely insulated from the tensions generated by economic competition. At a minimum the United States will continue to demand that its more prosperous allies, especially Japan and South Korea, assume more of the economic burdens of the common defense. At a maximum it is conceivable that disputes over economic matters could cause serious strains in the alliances between the United States and such countries as Japan, South Korea, and Thailand.

Nonetheless, although economic competition has complicated some friendly relationships in the Pacific Rim, it has also begun to unfreeze some previously hostile ones. As countries seek to maintain or improve their economic positions by finding new markets, new sources of capital, or new sources of technology, they are starting to look toward countries with whom their political relationships have previously been strained. Prominent examples include the expansion of economic relations between South Korea and China, China and Taiwan, Taiwan and Vietnam, Vietnam and Thailand, Indonesia and China, and South Korea and the Soviet Union. Economic factors are also at work in the reduction of tensions between China and the Soviet Union and may ultimately produce a partial reconciliation between the Soviet Union and Japan. In all these instances, common economic interests are beginning to catalyze an improvement in political ties.

Thus, as a result of the economic dynamism of the Asia–Pacific region, countries that Americans once regarded purely as military allies are increasingly regarded as economic competitors. Indeed, recent public opinion polls indicate that Americans now regard the economic competition of our allies as a more important threat to American security than the military challenge posed by traditional adversaries. At the same time, countries that once viewed each other as strategic rivals are finding advantage in closer economic interaction. Economic relations are beginning to grow between countries whose diplomatic relations are hostile, strained, or nonexistent. All this implies a much greater fluidity in strategic relations in the Asia–Pacific region, in that alliances are being challenged by economic tensions and confrontations are being eased by economic cooperation.

Trends toward Multipolarity

Together, the more conciliatory policy of the Soviet Union and the economic dynamism of the Pacific Basin are contributing to a fourth aspect of the regional security environment: the greater diffusion of military power and strategic influence in the area. This trend is often summarized in the term *multipolarity*. That word is misleading if it is taken literally to mean the emergence of additional superpowers with a combination of military and economic resources rivalling the Soviet Union or the United States. But it is appropriate if it is defined more loosely as the rise of a number of middle-level powers that are increasingly independent in orientation and assertive in policy. The emergence of such nations is leavening the bipolarity that had been characteristic of Asia through most of the 1970s.

The emerging middle-level powers of Asia fall into two categories. First, there are the large nonaligned countries, such as India, Indonesia, and China. These nations have traditionally conducted independent foreign policies, even though China had developed fairly extensive security ties with the United States before the events in Tiananmen Square in 1989, and India has maintained close strategic relations with the Soviet Union. Second, there are an even larger number of countries that, although formally allied with one of the superpowers, are increasingly defining their interests and formulating their foreign policies independently of their patrons. Examples include Japan, South Korea, Thailand, and Vietnam. Both sets of nations—aligned and nonaligned—have the economic resources, military potential, and political ambition to play more active roles on the international stage in the years ahead.

The ambitions of these middle-level powers vary. Some seem interested in establishing greater influence, or even in gaining paramountcy, in the region immediately surrounding them. India, through its annexation of Sikkim, its role in the division of Pakistan, and its interventions in Sri Lanka and the Maldive Islands, is clearly seeking a dominant position in South Asia and the Indian Ocean. Since the early 1950s Hanoi has attempted to create a confederation of Indochinese states oriented toward Vietnam. Thailand, with the most dynamic economy in ASEAN, now seems eager to expand its economic position in continental Southeast Asia, and perhaps its political role as well. Indonesia has long had the ambition to serve as the leader and spokesman for the rest of non-Communist Southeast Asia.

Some of Asia's middle-level powers are interested in playing more active roles in regional and global groupings and institutions. China and India continue to act as leaders of the Third World and the Nonaligned Movement, attempting to promote the formation of a new international economic order. Japan, South Korea, and Australia have been at the forefront of efforts to create some form of Pacific Community. Japan is trying to identify ways in which it can take a more active part in managing the international economy, providing more development assistance to the Third World, and promoting international stability. China seeks recognition as a major participant in the resolution of regional issues in both Asia and the Middle East, in part by an aggressive program of arms sales to both regions. Even Taiwan is attempting to regain membership in a variety of regional and global economic organizations.

As these middle-level powers rise to greater prominence, the strategic environment of the Asia–Pacific region will increasingly be shaped by the rivalries among them. Some of these tensions are the result of territorial conflicts of varying degrees of severity. China and Vietnam, China and India, and China and the Soviet Union all have disputed borders that have been the scene of direct military confrontation in the past. China, Vietnam, Malaysia, the Philippines, and Indonesia have competing claims to islands in the South China Sea. Further north, China and Japan dispute the ownership of the Senkaku (or Diaoyutai) islands northeast of Taiwan, and Beijing and Seoul differ over the division of undersea resources in the Yellow Sea. The territorial conflict over four groups of islands off Hokkaido is the most serious obstacle to the improvement of relations between Japan and the Soviet Union.

Ethnic tensions could also cause problems between these middle-level powers. Rivalry between Hindus and Moslems lies at the heart of the continuing tensions between India and Pakistan, and rivalry between Malays and Chinese could complicate Singapore's relations with either Indonesia or Malaysia. The prominent role of overseas Chinese in many Southeast Asian countries, both Communist and non-Communist, has been a significant irritant in those nations' relationships with China. The discrimination encountered by Koreans in Japan could easily become an issue in Tokyo's relationship with Seoul.

The overlapping geopolitical ambitions of several of these middle-level powers could also produce tension and instability in the years ahead. China and India compete for influence in South Asia; China, Vietnam, and Thailand in Indochina; Indonesia and China in Southeast Asia; and Australia and Indonesia in Melanesia. More broadly, there is concern throughout the Pacific Rim about the rising military power of both China and Japan and about the possibility that either Tokyo or Beijing could assert a claim to leadership in East Asia. Growing Japanese investment and foreign assistance, although welcomed in some respects, also generates worry about the rise of Japanese political influence.

The emerging multipolarity in East Asia has several significant implications for the United States. If Soviet policy toward the region remains flexible and accommodating, and if Gorbachev continues to freeze or reduce Soviet force levels in the Far East, then the other countries of the Pacific Rim will begin to pay relatively less attention to the Soviet threat and will start to place more emphasis on the challenges posed to them by the other middle-level powers of Asia. Japan may be concerned as much with China as with the Soviet Union. China may focus less on the Soviet Union and more on the potential threats posed by India and Japan. And Southeast Asian nations may worry more about the intentions of Beijing and Tokyo than about those of Hanoi or Moscow.

This, in turn, will complicate American alliances in several respects. Common opposition to Soviet expansion may be a less compelling basis for America's strategic relationships in the 1990s than it was in the 1970s and early 1980s. It will therefore be necessary to find a new foundation for the alliances linking the United States with Japan, the Philippines, and Thailand, and for the relationship between China and the United States. Conversely, the United States may find it increasingly difficult to maintain close security ties with pairs of countries that come to regard each other as their principal strategic rivals. The emergence of serious tensions between China and Japan, between Australia and Indonesia, or between India and China would be a significant factor complicating American strategy in the Asia–Pacific region.

On the other hand, the emerging diffusion of power in East Asia also provides a new rationale for the deployment of American forces in the region. To be sure, American forces will still be needed to perform their traditional missions of counterbalancing the Soviet Union and defending Japan and South Korea. Increasingly, they will also help maintain a broader balance of power in the region, both by deterring aggression by regional powers and by obviating the need for other nations to enlarge their own military forces. The role of the United States will gradually evolve from that of the architect of containment in a bipolar region to the guarantor of balance in a more multipolar Asia.

ISSUES FOR AMERICAN POLICY

These four elements raise a number of more specific issues for American policy toward the Asia–Pacific region in the 1990s. These include:

- The continuing need for the forward deployment of American forces.
- The successful management of American alliances with our partners in the Pacific Rim.
- The development of a cost-effective military strategy for the area.
- The pursuit of appropriate arms-control measures for the Asia–Pacific region.
- The development of an effective diplomatic strategy for dealing with a more multipolar Asia.

Although this essay cannot provide an exhaustive discussion of each of these five issues, it can suggest some of the principal questions that a more comprehensive analysis would need to address.

Forward Deployments

The changing security environment in East Asia, the economic prosperity of most of the region, and the financial constraints facing the United States inevitably raise the question of whether the United States should continue to maintain a sizeable military presence on the Western side of the Pacific Rim. The answer remains affirmative, but the justification for those forward deployments is gradually changing, and the size and location of American forces require reexamination.

In the past, American deployments were needed to help defend South Korea against the North, Japan against the Soviet Union, and Taiwan against mainland China. Today several of those threats appear to be subsiding, and each of America's allies is able to bear a greater responsibility for its own defense. But American deployments remain necessary in fulfilling residual commitments to the security of our allies in the Western Pacific. They also enable the United States to project military force effectively into the Indian Ocean and the Persian Gulf. Perhaps most important, as suggested above, they help deter the regional arms races in the region that would almost certainly occur in the vacuum formed by an American withdrawal.

Moreover, the forward deployment of American forces in the Western Pacific can be maintained at an acceptable cost, both diplomatically and financially. The utility of American forces in preventing the rapid growth of Japanese military power is widely acknowledged by analysts in both China and the Soviet Union. Despite opposition from some radical elements in the two countries, American forces are still welcomed in both Japan and South Korea. The role of the United States in dampening regional arms races and in maintaining a regional balance of power is also accepted by most nations of Southeast Asia. These responsibilities are presently being fulfilled at a cost of about one-tenth of the American conventional military budget—a remarkably small ratio given the importance of the region to the United States.

But where should American forces be stationed? American deployments in Japan, the cornerstone of our force structure in the Northwest Pacific, seem secure, although the environmental impact of our bases may occasionally be controversial. American installations in the Philippines, in contrast, are much more problematic. It now appears likely that American forces will be permitted to remain in the country after the treaty governing them expires in 1991, but only on the condition that they ultimately be removed from the Philippines. Given the importance of forward deployments in the Southwest Pacific to the American ability to project force into Southeast Asia, the Indian Ocean, and the Persian Gulf, it would be desirable for the United States to accept the cost of relocating the bases elsewhere in the region if they cannot be kept in the Philippines.

American deployments in Korea are also subject to reconsideration. It is now generally accepted that the growth of the South Korean economy will gradually permit Seoul to assume greater responsibility for its own defense and will simultaneously allow the United States to reduce the size of its ground forces in the country. A plan for cutting American troop strength in South Korea by approximately 10 percent was recently announced in Washington.[3] Further reductions would necessarily also entail the withdrawal of American nuclear forces from the Korean peninsula. In the meantime the planned relocation of American forces away from downtown Seoul will help remove an emotional irritant in Korean–American relations.

Alliance Management

A second issue facing the United States in the 1990s will be the management of the tensions that are already arising in our alliances in East Asia. The increasing prosperity of most of our allies and the financial difficulties of the United States make the notion of burden sharing a most attractive proposition. But while it is appropriate for some of our allies, particularly South Korea and Japan, to assume greater responsibility for their own defense, we also need to appreciate some of the dilemmas that are inherent as we attempt to share the strategic burden more equitably.

First, we need to be aware of the enormous suspicions that are already being created by Japanese rearmament. Most of these apprehensions are expressed by other nations in the region, many of whom were victims of Japanese aggression during World War II. But, as the FSX issue has suggested, there is a growing concern in the United States that the development of the Japanese defense industry will serve ultimately to improve Japanese competitiveness in advanced civilian technology. This suggests that in many respects it would be more acceptable for Tokyo to bear a greater share of the financial cost of stationing American troops in Japan than to significantly increase its own military strength or production capability. But it remains to be seen whether either Japanese or Americans will be comfortable, over the long run, with a relationship in which American forces serve, in effect, as mercenaries for a foreign power.

Second, it will be necessary for the United States to manage its economic relations with East Asia wisely, so as to prevent economic problems from undermining its security ties with its allies. It is appropriate, as well as politically necessary, for Washington to press for greater access to foreign markets. But this general approach needs to be qualified in several ways. The United States must keep its own markets open if it is to demand greater access to markets elsewhere.

If Washington is to press for changes in the macroeconomic policies of its trading partners, it will have to be able to demonstrate that it is taking effective measures to address its own problems, including lagging productivity, inadequate levels of savings, and high budget deficits. Finally, the United States will have to consider carefully whether the economic benefits gained from forcing access to especially sensitive sectors of Asian markets—especially in agriculture—will be worth the possible damage to our political relationships in the region.

A third dilemma concerns the way in which a common defense policy is formulated within American alliances. If the United States is going to share the burden with its allies, it will also have to share the power to determine common alliance policy. We cannot expect other countries to pay more for a defense strategy or for military deployments that are decided on entirely in the United States. This is particularly true at a time when the strategic environment in Asia is changing rapidly, and when defense policy will have to be modified to respond to new challenges and opportunities. The adjustments in policy—toward the Soviet Union, toward China, toward regional conflicts in Korea and Indochina—will have to be devised and implemented through mechanisms that are much more consultative and multilateral than has been the case in the past. One specific example is the need to reconsider the division of responsibility between Korean and American officers in the Combined Forces Command on the Korean peninsula.

Military Strategy

One of the issues that will need to be addressed through these new consultative mechanisms is the most appropriate American military strategy to deal with the strategic environment in East Asia in the 1990s. In the 1980s the Reagan administration's strategy in the Pacific Rim was strongly influenced by the doctrine of horizontal escalation, espoused by some leading civilian and military officials in the U.S. Navy. Under that doctrine, the United States would have developed the capability to attack Soviet military installations in the Far East in the event that a conventional war between two superpowers broke out in Europe or the Middle East. A strategy of horizontal escalation justified the navy's plan for a substantial increase in the strength of the Seventh Fleet.

Even in the early years of the Reagan administration, objections were raised to the strategy of horizontal escalation. Some challenges focused on the effectiveness of the strategy, questioning whether conventional forces could successfully undertake the missions and responsibilities assigned to them without suffering unacceptable losses. Others noted the high cost of the naval and air forces required if the strategy was to have any chance of success. Still other critics pointed out the damage that a strategy of horizontal escalation might do to American alliances and prestige in the Western Pacific. Those countries with

American bases might be subject to retaliatory or preemptive action from the Soviet Union and might therefore find their own security reduced rather than enhanced by their strategic relationship with the United States. Moreover, the strategy of horizontal escalation implied that it would be the United States, not the Soviet Union, which would bear the responsibility for extending into the Asia–Pacific theater a conventional war that had begun in some other region.

Now, with new leaderships in office in both Moscow and Washington, the criticisms of the strategy of horizontal escalation appear even more compelling. The receding Soviet threat to the region makes it even more difficult to justify so expensive a strategy. The constraints on the American defense budget are even more severe. And it is even less likely that the concept of horizontal escalation would attract the broad international support necessary at a time when, as argued above, our defense policy in Asia must be devised through more consultative mechanisms.

A challenge for the 1990s, therefore, will be to redefine American military strategy in Asia to meet the new context of emerging multipolarity, financial stringency, and a seemingly less expansionist Soviet Union. Such a strategy would place less stress on an offensive capability against the Soviet Union and greater emphasis on the defense of American allies and sea lines of communication, and on the maintenance of flexible rapid deployment forces to meet a broad range of contingencies.

Arms Control and Regional Disputes

A fourth challenge to the United States will be to respond to the arms-control initiatives for the Asia–Pacific region that have been proposed by Gorbachev in Vladivostok, Krasnoyarsk, and Beijing since 1986. The Soviet initiatives to date fall into two broad categories. First, there are a number of measures to be undertaken unilaterally by Moscow, including

- a freeze on Soviet nuclear weapons deployed in Asia, apparently on the condition that this step be followed by similar actions by other nuclear-weapons states.
- the withdrawal of all Soviet forces from Mongolia, with one division removed in 1988, two tank divisions and the entire air force contingent withdrawn in 1989–1990, and the remaining division to be removed shortly thereafter. Some of these forces will be redeployed elsewhere, rather than completely demobilized.
- a reduction of forces in the Far East by some 120,000 troops, together with the removal of some 80,000 additional troops from more westerly parts of Soviet Asia, both to occur in 1989–1990. This is to involve the reduction of twelve army divisions, eleven air force regiments, and sixteen combat ships.

A second set of Soviet initiatives involves proposals for multilateral arms-control agreements covering the Western Pacific, including

- consultations with the main naval powers of the region on a freeze of naval forces in the Western Pacific.
- discussions among the Soviet Union, China, Japan, and the two Koreas—apparently not involving the United States—on a freeze and subsequent reduction of naval and air forces in the areas where their coasts converge.
- the closing of the Soviet naval station in Cam Ranh Bay in exchange for the elimination of all American military bases in the Philippines.
- measures to prevent incidents at sea and in international airspace in the Asia–Pacific region.
- an international conference on making the Indian Ocean a zone of peace.
- creation of nuclear-free zones on the Korean peninsula and Southeast Asia.
- negotiations with China on a mutual and balanced reduction of forces along the Sino-Soviet border.
- consultations among the Soviet Union, China, and the United States on the creation of a negotiating mechanism on the security of the Asia–Pacific region. This last proposal represents the modification of the earlier initiative, put forward at Vladivostok in 1986, calling for a Helsinki-style conference on mutual security in the Asia–Pacific region.

Thus far, the United States has made a minimal response to these Soviet arms-control initiatives. An op-ed essay in the *New York Times* in October 1988, which was coauthored by an assistant secretary of state and an assistant secretary of defense, and which summarily rejected the Soviet proposals as entirely one-sided,[4] remains the basic statement of American policy on this matter. But many analysts, both in the United States and in Asia, worry that such a reaction is overly negative and excessively passive. Washington cannot afford, in their view, to allow the Soviet Union to seize and hold the initiative on regional arms-control issues without putting forward proposals of its own that can gain widespread support in the region.

A more creative American posture might contain three elements. First, we must vigorously respond to the Soviet proposals that are unacceptably one-sided, such as the attempt to equate the Soviet naval base at Cam Ranh Bay with the entire American base structure in the Philippines and the various proposals for nuclear-free zones in the Indian Ocean, Southeast Asia, and Korea. We could do so not simply by rejecting unreasonable Soviet initiatives, but also by putting forward some of our own proposals that would be disadvantageous to the Soviet Union, such as a nuclear-free zone that would include portions of the Soviet Far

East. Such initiatives would be designed to reveal the strategic asymmetries between the Soviet and American positions in the Western Pacific.

Second, we need to respond more favorably to the items on the Gorbachev agenda that are more promising. These include, most prominently, the idea of multilateral negotiations on the prevention of incidents at sea or in international airspace, which could well build upon existing bilateral agreements between the Soviet Union and United States on preventing incidents at sea and on the multilateral agreements reached after the tragic destruction of KAL Flight 007 regulating international airspace in the Northwest Pacific. Some of Gorbachev's proposals for a restriction of nuclear forces, naval forces, and air forces in the Western Pacific are also worthy of consideration, although it may be more practical to achieve those goals through an informal process of mutual restraint than by formal arms-control negotiations.

Finally, the United States should continue to press forward with its own agenda for reducing tensions in the Asia–Pacific region. Traditionally the United States has placed greater emphasis on resolving regional disputes than on formal arms-control negotiations in East Asia. This is based upon the assessment, alluded to above, that formal arms-control negotiations have proven to be difficult and protracted even in a bilateral context and can be expected to be even more lengthy and complex in a multilateral arena. The American approach also reflects the conviction that both military conflict and arms races in the Asia–Pacific region are symptoms of underlying local disputes and that a more fruitful strategy of tension reduction in the region may, therefore, be to address these more fundamental issues, including Cambodia, Korea, the Taiwan Straits, the Northern Territories, and the Sino-Indian frontier.

As already suggested the most appropriate American strategy for dealing with regional disputes is to encourage contacts between the parties directly involved. In Cambodia, negotiations are well underway toward a comprehensive political solution. In Korea, there are good prospects for an expansion of economic relations, for the installation of confidence-building measures along the demilitarized zone, and for political dialogue between the governments in Seoul and Pyongyang. In the Taiwan Straits, the expansion of economic and cultural relations could ultimately be supplemented by unofficial dialogue on political questions. Japan and the Soviet Union have begun to discuss the fate of the Northern Territories, and Peking and New Delhi have agreed to establish a vice-ministerial-level working group on their disputed border. Obviously, the American involvement in these particular conflicts will vary from case to case. In the broadest sense, however, the role of the United States should be to help create the international climate to facilitate a reduction of tensions and, where appropriate, to serve as a guarantor of the solutions that are arrived at by the parties directly involved.

Diplomatic Strategy

The last challenge for American policy in Asia will be to construct an effective diplomatic strategy for dealing with the steady diffusion of power in the Pacific Rim. The United States must recognize that the Soviet Union is not the only threat to regional stability and that the American diplomatic posture can no longer be based exclusively on the creation of a united front against Soviet expansion. Instead, the overall U.S. strategy should be to help maintain a balance of power among all the nations of the region, promoting accommodation where possible, but resisting the efforts of any country to threaten the security of its neighbors.

Such an approach could build upon three more specific elements. First, it is necessary to expand bilateral dialogue, on both the official and unofficial levels, with allies and potential adversaries alike. We should not expect to reach complete consensus or gain automatic support for American initiatives, but we should acknowledge instead that diverse opinions will be inevitable. The aim should be to identify areas of common interest and perspective and to develop policies that can build effectively upon them. As part of this approach, the United States should gradually develop contacts with countries such as Vietnam and North Korea, as well as expand its dialogue with all other major actors in the region.

Second, our bilateral consultations need to be supplemented, where feasible, by multilateral forums on both economic and security issues. It may not be possible to construct a single international economic institution in Asia, comparable to the European Community. But the PECC already provides a mechanism with near-universal membership for unofficial and semiofficial consultations on economic matters. Subregional organizations, such as ASEAN, the South Pacific Forum, and the South Asian Association for Regional Cooperation, offer opportunities for formal consultations on economic issues among smaller groups of nations. And the conference on Asia-Pacific Economic Cooperation (APEC), which first met in Canberra in November 1989, brings together official representatives from ASEAN, South Korea, and the developed states of the Pacific Rim for discussions of trade and investment matters.

Given their sensitivity, security questions may require more carefully tailored forums. But it may now be possible to envision a regular unofficial dialogue among academic specialists on international security from a wide range of Asia–Pacific nations. Such a forum would be analogous to the Pacific Trade and Development Conference (PAFTAD), a group of scholars of Asian economic issues that has met periodically since 1968. If such an organization proved successful, it might then be possible to add participants from the diplomatic and military communities to the dialogue, so as to create a counterpart to the PECC for security issues. Its aim would be to minimize misunderstanding and misperception, moderate regional rivalries, and seek cooperative solutions to common problems.

Finally, the United States must take the increasingly multipolar character of East Asia into account when formulating its policies on burden sharing and arms transfers. The aim can no longer be to strengthen every possible participant in a united front against the Soviet Union. Instead, we must understand the ways in which the rapid development of the military power of China, Japan, India, or any other regional leader could prove destabilizing in a more complex regional balance of power.

CONCLUSIONS

Perhaps the greatest challenge confronting the United States in Asia is the need to adjust its policy in the absence of crisis. From many perspectives the present situation in the Asia–Pacific region is good for the United States. The American economy remains reasonably strong, our trade deficits have been gradually declining, the Soviet Union has adopted more accommodating policies, there are new prospects for resolving major regional disputes, and the United States has good relations with most of the emerging powers of the region.

Given such happy circumstances, it is tempting for busy policymakers to continue on their present course, often invoking the time-honored adage, "If it ain't broke, don't fix it." On reflection, however, this metaphor proves to be woefully anachronistic. Such an approach may have been appropriate for the technology of the nineteenth century, where simple machinery usually caused little damage when it failed. But it is much less apt for the more complicated technology of the late twentieth century, where the malfunction of equipment or institutions can have catastrophic consequences. In a complex world, regular upkeep is vastly preferable to emergency repair.

There is, in other words, the need for preventive maintenance for our Asia policies—not to respond to any present crisis but to prevent a future one. As this essay has suggested, this type of preventive maintenance will involve creative and timely responses to the new policies of the Soviet Union, the economic dynamism of the region, the change in the relative power of the United States, the gradual diffusion of influence in Asia, and the new prospects for resolving regional disputes. We can no longer base our policies on old assumptions: that the Soviet Union is the only threat to the stability of the region, that the maintenance of an anti-Soviet united front can be the cornerstone of American strategic doctrine in Asia, and that the United States can be the dominant partner in all its alliances.

As the discussion of concrete issues has implied, there is no need for sudden or wrenching change in American policy. But there is the necessity for deliberate, sustained adjustment to meet new challenges and to seize new opportunities. If we fail to do so, we will find that our policies will gradually diverge from the realities of the emerging Asia.

NOTES AND REFERENCES

1. This essay was prepared for delivery as a lecture on December 9, 1988, and was slightly revised and expanded for inclusion in this volume. It draws, in part, on two of the author's previous essays on American relations with East Asia: Harry Harding, "The American Strategy in the Far East," in Kjeld Erik Brodsgaard, *East Asian Security and Foreign Policy in the 1980s*, Copenhagen Papers in East and Southeast Asian Studies, no. 2/88 (Copenhagen: Center for East and Southeast Asian Studies, University of Copenhagen, 1988), pp. 81–97; and Harry Harding and Edward J. Lincoln, "The East Asian Laboratory," in John D. Steinbruner, ed., *Restructuring American Foreign Policy* (Washington, DC: The Brookings Institution, 1988), pp. 185–220.
2. For the Vladivostok speech, see Moscow Television, July 28, 1986, in *Foreign Broadcast Information Service Daily Report: Soviet Union*, July 29, 1986, pp. R1–R20; for the Krasnoyarsk address, see TASS, September 17, 1988; for Gorbachev's speech in Beijing, see Moscow Television, May 17, 1989, in *Foreign Broadcast Information Service Daily Report: China*, May 18, 1989, pp. 12–18.
3. *The New York Times*, February 16, 1990.
4. Gaston J. Sigur and Richard L. Armitage, "To Play in Asia, Moscow Has to Pay," *The New York Times*, October 2, 1988.

9

Why the Third World Matters[1]

Steven R. David

The Third World has been and will remain central to U.S. interests.[2] The risks of superpower confrontation, the use of nuclear weapons, and American (and Soviet) soldiers engaging in combat are all greater in the Third World than in Europe or Japan. Economic disaster to the United States and its allies is more likely to arise from developments in the Third World than anywhere else. It is in the Third World that the broader receptivity to American goals and values will be determined. In short, the instability and ferment characteristic of the Third World will continue to engage American interests with an urgency and unpredictability unmatched by its so-called vital allies.

The apparent dismantling of the Soviet empire in Eastern Europe, along with domestic unrest in the USSR itself, has enhanced at least the relative importance of the Third World to the United States. The two primary threats to American interests—a superpower nuclear war and a Soviet invasion of Western Europe—are now less likely to arise than at any time in the past forty years. Meanwhile, the power of many Third World states, some deeply hostile to the United States, continues to grow at a rapid rate. Failing to concentrate on addressing the actual and potential threats they pose would seemingly be the height of irresponsibility for American policymakers.

Nevertheless, there is a growing literature arguing that the United States exaggerates the importance of the Third World to its interests. Although disputes exist among adherents of this view, there is agreement on several fundamental points. Proponents of this view claim that the Third World does not pose a threat to the vital interests of the United States (defined as the preservation of American

---

Steven R. David • Department of Political Science, The Johns Hopkins University, Baltimore, Maryland 21218.

security, economic well-being, and core values). Any threat to the limited U.S. interests in the Third World is not so much from the Soviet Union or radical revolutionaries, but from misguided American policies. By pursuing an aggressive, activist policy the United States will drive Third World regimes into the arms of the Soviets, thus bringing about the outcome it is seeking to avoid. The best way for the United States to maintain its interests in the Third World is to pursue an accommodationist policy that recognizes its inability to control Third World developments. Above all, the United States must avoid the trap of squandering scarce resources on peripheral Third World interests while truly vital concerns receive inadequate attention.[3]

These analysts, whom I will call *hyper-realists*, have provided a service in forcing the explicit consideration of just what is and is not important in the Third World, and how American policy can best secure U.S. interests.[4] Moreover, the hyper-realists are correct in arguing that the United States must give top priority to its own protection and the protection of its Western European and Japanese allies. But this does not require an abandonment of American interests and commitments in the Third World that the hyper-realist approach would bring about, all the more so because an engaged American policy in the Third World can be carried out at a reasonable cost without requiring the protracted use of American personnel abroad. We cannot allow fears of threats that almost certainly will never materialize to prevent us from dealing with Third World threats that already exist, as well as the far more serious challenges to American interests that are likely to develop in the future.

My argument is put forth in three parts. First, I will review some of the principal assertions of those who believe the United States exaggerates the importance of the Third World and should play a less active role in trying to determine the outcome of developments there. Second, I will explain why their assertions are incorrect or are themselves exaggerated, and why the United States needs to be actively involved in attempting to influence the course of Third World events. I will then briefly discuss the approach the United States ought to take to the Third World.

THIRD WORLD THREATS TO AMERICAN INTERESTS

The Strategic–Military Threat Posed by the Third World

A major tenet of the hyper-realist argument is that the Third World poses only a negligible threat to U.S. security interests. The hyper-realists emphasize that Mikhail Gorbachev recognizes the limitations of the Soviet appeal in the Third World and is disappointed by the high costs and meager gains of previous Soviet Third World policy. The Soviet decision to withdraw from Afghanistan and its encouragement of a Cuban withdrawal from Angola are seen as tangible support for their view that Gorbachev is serious about downgrading Soviet involvement in the Third World as part of his new political thinking.

Buttressing the notion of a lessened Soviet interest in the Third World are the momentous events that have shaken Moscow in the late 1980s and early 1990s. At home the Soviet Union confronts secessionist demands and escalating unrest from Lithuania, Estonia, Latvia, Azerbaijan, and Armenia. These difficulties, combined with the revolutionary decision to abandon the monopoly of power of the Communist party, are understandably more important to the Kremlin than efforts to extend Soviet influence in the Third World. Insofar as the Soviet Union can deal with foreign-policy matters, it must confront the implications of the collapse of its position in Eastern Europe and the reunification of Germany before turning to the less pressing issues presented by the Third World.

Even if the USSR chooses to once again attempt to extend its influence in the Third World, the hyper-realists assert, there would be no cause for American concern. The United States need not fear Soviet control of Third World groups and countries because in the hyper-realist view the USSR will never achieve the kind of domination of the Third World necessary to enhance Soviet power. In part, this belief rests on the hyper-realist judgment that the USSR has historically not done well in the Third World and that Soviet losses (e.g., Egypt) have outweighed their gains. More important, the hyper-realists argue that the USSR lacks the capability and will to make significant inroads in the Third World in the future.

For the hyper-realists, the USSR has little to offer Third World states. Economically, it accounts for only 2 percent of Third World trade while contributing negligible amounts of aid. With decolonization virtually complete, the appeal of Soviet ideology has diminished. Even in the area of military assistance—clearly the USSR's strong suit—Moscow does not have a durable instrument of influence, as evidenced by its setbacks in states such as Egypt, Sudan, and Somalia, where the Soviet Union had been the chief arms supplier.

Should the Soviets adopt an activist policy in the Third World, the hyper-realists reassure us that nationalism and better indigenous security forces are making Third World regimes increasingly difficult to control. Barry Posen and Stephen Van Evera argue that "ultimately this serves American interests, since America's chief purpose is not to establish world dominion but rather to keep the world free from Soviet dominion."[5] Similarly, the tendency of states to balance (by aligning against an aggressive power) will prevent the USSR from reaping the benefits of a threatening posture. Nor, say the hyper-realists, need the United States fear that successful Communist revolutions can be exported to other Third World states.

Finally, the hyper-realists assert, even if the Soviets succeed in establishing bases throughout the Third World, America's security will not be threatened. The United States can mitigate the impact of Soviet bases by, for example, deploying new forces. More important, Soviet bases in the Third World do not matter because of the existence of nuclear weapons. Inasmuch as "any direct prolonged superpower war would likely become nuclear," fear of Soviet bases in "strategic" locations such as the Gulf of Mexico is an example of outdated thinking that employs a pre-1945 mentality to a world transformed by the nuclear revolution.[6] In sum, the USSR is not likely to focus its efforts on the Third World, would not be successful if it did, and even if successful, would not seriously challenge American security interests.

The Threat to American Economic Interests Posed by the Third World

The hyper-realist view that the Third World does not pose a significant economic threat to the United States rests on two arguments. First, it asserts, the Third World lacks the ability to hurt the United States through economic means. The aggregate gross national product (GNP) of the entire Third World is less than three-quarters that of Western Europe and only one-half that of Western Europe plus Japan. Thus, Soviet domination of the Third World would not grant Moscow the military potential that control of more important industrial regions (i.e., Europe and Japan) would give them. Even the control of raw materials by Third World states is not cause for concern because these states are supposedly unable to form cartels, such as the Organization of Petroleum Exporting Countries (OPEC), that could embargo the West. Should a cutoff nevertheless occur, the hyper-realists allege, alternative suppliers and stockpiles would mitigate its effects.

Second, even when it is acknowledged that the Third World does possess significant economic instruments, the hyper-realists assert that self-interest dictates that they will not use them. Regardless of their ideology or degree of pro-Soviet orientation, all Third World states, it is argued, recognize the necessity to participate in the Western-dominated international economy. As Feinberg writes, "Whether they are neoliberals, populists, social democrats, or Marxists, most Third-World leaders desperately want to participate in this new international system."[7]

Thus, insofar as American trade and investment with the Third World are important to the United States, the hyper-realists allege, the United States has little to fear so long as it does not needlessly antagonize Third World states by trying to determine the character of their regimes. Similarly, the hyper-realists argue that the United States need not fear a boycott by strategic mineral-producing states because such an action would hurt their own fragile economies far more than it would the United States.

The Threat to American Political–Ideological Interests Posed by the Third World

For the hyper-realists, American political–ideological interests (e.g., maintaining freedom and democracy) in the Third World are of only marginal concern. While it would be gratifying if Third World states embraced American values, it is not necessary that they do so. Few Americans are even aware of Third World political systems. If Third World leaders opt for Marxism, the hyper-realists assert, it would have little or no effect on the strength of the American political system. American democracy simply does not depend on the existence of democracies elsewhere, particularly in the Third World. The only threat to our values posed by the Third World is the tendency of the United States to claim that those values are at stake in Third World conflicts, when in fact they are not.

The Hyper-Realist Approach to the Third World

The hyper-realists see America's chief concerns in the world as protecting its allies in Europe and Japan, and preserving the security of the United States. In both of these areas the Third World (with the possible exception of the Persian Gulf) plays a marginal role at best. Those few interests that the United States does maintain in the Third World are not seriously threatened and can be easily safeguarded without our becoming involved in Third World conflicts.

The United States has more to lose in the Third World, they argue, by acting than by not acting. The inability of the United States to control events in the Third World is not a cause for concern to the hyper-realists because they believe that regional powers and nationalism, rather than the Soviet Union, will fill any gap left by the decline of American influence. So long as revolutionary states do not launch armed attacks on the truly vital interests of the United States, a policy of nonintervention will prove more beneficial to American interests than attempts to topple unfriendly regimes. As for the prospect of the USSR's extending its influence in the Third World, "the best response to a Soviet inroad may be to give the Soviets time to make their own mistakes."[8]

Most important, as Stephen Walt argues, because we live in a *balancing* world (in which states align against an aggressive power) rather than a *bandwagoning* world (in which states align toward an aggressive power), forceful policies by the USSR (or anyone else) will not succeed in attracting allies. Therefore, the United States (a status quo power) can be complacent about most international developments.[9]

The hyper-realists also emphasize the limits of U.S. power and argue that the United States must avoid wasting scarce resources in the Third World. This means spending far less than the United States is now doing for Third World contingencies. The hyper-realists want the United States to scale down its Third World interventionary forces, cut back on its commitments throughout the Third World, and lessen its reliance on security instruments to accomplish American objectives. These steps will not only save money, they say, but also lessen the prospect of another disastrous intervention such as occurred in Vietnam.

RESPONDING TO THE HYPER-REALISTS

Before challenging the specifics of this critique, it will be useful to confront the thrust of the argument, namely, that the Third World contains few, if any, threats to the vital interests of the United States. The hyper-realists err by failing to define what they mean by *vital* and/or by an undifferentiated lumping together of all Third World states. It is unarguably true that the Third World threatens few vital interests if vital means the preservation of American security, economic well-being, and core values. But such a standard also means that the United States faces few threats to its vital interests in the rest of the world. Many of the key arguments used to demonstrate the lack of importance of the Third World could be employed to justify a policy of noninvolvement in Europe. Such a result is especially disturbing in that it does not appear that any of the hyper-realists would seek this outcome.

Militarily the arguments of the hyper-realists would also support an American withdrawal from Western Europe. Because we supposedly live in a world of balancing, the Western Europeans (whose combined GNP is larger than that of the USSR) would simply take it on themselves to confront the Soviet Union. Nor, according to the logic of some of the hyper-realists, would the United States have to fear Western Europe's industrial power falling to the Soviets. If nuclear war makes protracted conventional war virtually impossible, Soviet control of Europe's industrial potential would not enhance the USSR's threat to American security because any major conflict would be long over before Europe's economic strength could play a role.

Economically the arguments of the hyper-realists suggest that the United States need not fear domination of Western Europe by the Soviet Union or radical governments. After all, if ideology does not determine trading partners, it does not matter what kind of regime is in power. With all governments seeking to participate in the international economy, a Communist Western Europe should be just as reliable and valuable a trading partner as the Western Europe of today. That a Communist Europe would trade freely with the United States and contribute little to the military threat posed against it also calls into question the view that industrial power is the best gauge of strategic importance.

In terms of the threat to American political–ideological interests, as conceived of by the hyper-realists, Europe is also expendable. If the survival of American democracy does not depend on the internal characteristics of other governments, it is difficult to label such characteristics as vital. To do so stretches the meaning of *vital* to such an extent as to cloud its meaning. For those who accept the principle that American democracy depends on democracy elsewhere, on the other hand, the existence of democratic governments in the Third World should be no less vital than their counterparts in Western Europe.

The point of this comparison with Europe is not that the Third World is as important as Europe or that policymakers should not make distinctions among interests; rather, the United States is fortunate in that its vital interests (the preservation of American security, economic well-being, and core values) are not under serious threat from any quarter. The hard choices come not in determining what is vital, but in deciding which nonvital interests are worthy of defending and how to do so. By using the standard of vital interests to direct American policy in the Third World (while applying it much more loosely to Europe), the hyper-realists are led to dismiss countries as unimportant simply because they are not vital. Such an approach will lead to isolationism whether or not that is the intention of its adherents.

Furthermore, it is misleading to assert that our interests in Europe are somehow vital while those in the Third World are not without making distinctions among individual countries. An outbreak of civil war in Mexico would threaten American interests more than a similar occurrence in Spain. The ascension to power of a Marxist–Leninist regime in South Korea would arguably be more damaging to the United States than a similar regime taking power in Portugal. Interests in the Third World, as elsewhere, must be judged on their individual merits. This prevents the dividing of the world between the vital interests of Europe and the nonvital interests of the Third World.

Broadly speaking, just how important are American interests in the Third World? First, it is difficult to identify in advance which interests and countries are likely to be important. When Secretary of State Dean Acheson excluded South Korea from the range of American vital interests, he did so because South Korea lacked intrinsic importance to the United States. The subsequent invasion by North Korea (probably encouraged by Acheson's action) demonstrated that countries of seemingly little significance can gain in importance when threatened (even indirectly) by Soviet power, thus calling into question America's credibility as an ally. In the 1960s, few predicted that Saudi Arabia would assume the significance it did scarcely a decade later. Chad is almost stereotypical of a Third World country lacking in importance to the United States. And yet, when it was threatened by Libya, Chad became important as a symbol of American resolve to help protect countries from Muammar Khadhafi expansionist designs. Whether the specific policies applied to these countries were justified is the subject of legitimate dispute, but it is impossible to determine a priori where American policy will next be engaged.

Second, the United States needs to devote more diplomatic and military attention to Third World contingencies, not because vital interests are at stake, but because U.S. interests are more likely to be threatened by what happens in the Third World. The United States has focused its efforts on interests of high intrinsic worth that confront relatively small risks: defending Western Europe and avoiding nuclear war. Third World interests cannot match these in importance, but there is a far greater probability that they will be threatened. When deciding how to spend the marginal dollar, the greater likelihood of risk to American interests from the Third World must be given weight.

The greater probability of Third World developments threatening American interests is enhanced by the tumultuous changes in the Soviet Union and Eastern Europe. The prospect of a Soviet-backed invasion of Western Europe (and Japan) or a U.S.–Soviet nuclear war has reached its lowest level in decades. At the same time, the threat to American interests from the Third World remains high and, in some cases, is growing. It makes little sense to ignore this Third World challenge because of fears of threats that are increasingly unlikely to emerge.

The United States must be especially concerned about the Third World, because that is where it stands the greatest chance of becoming embroiled in some violent conflict. We live in an age in which wars between and within developed states have virtually ceased to exist. There are many reasons given for this unprecedented phenomenon, including fear of nuclear war, economic entanglement, desire to avoid another World War II, and a growing acceptance of Western liberal democracy. The disagreements about why war has ended outside the Third World should not be allowed to obscure the recognition that conflict continues unabated inside the Third World. Nearly all the armed conflicts since World War II (including civil wars) have occurred in the Third World. All the wars in which the United States has been involved have been wars in the Third World. Insofar as large-scale conflict has the power to bring about U.S. involvement and threaten American allies, the mitigation of such conflict is in American interests.

Moreover, even if by some objective standard, reasons can be offered why the United States should stay out of Third World conflicts, domestic politics ensure that the potential of American embroilment will remain. Whether objectively correct or not, the fact that the American people worry more about threats from the Third World than from any other source makes dealing with those threats a matter of political survival for any administration. It can be argued, for example, that terrorism does not play a central role in threatening the interests of the United States. Nevertheless, when terrorism becomes an overriding concern to the American public it becomes important to those politicians seeking or wishing to remain in office. Objectively, the Third World might not be important enough to justify the worsening of U.S.–Soviet relations or the derailing of arms control treaties, but it has historically done just that. Objectively, Third World interests might not warrant superpower interventions (such as in Vietnam and Afghanistan) and nuclear alerts (such as occurred during the 1973 Middle East war), but experience demonstrates their potential for recurrence.

The hyper-realists, of course, put forth their arguments in an attempt to prevent these very developments from arising in the future. Nevertheless, they should be realistic enough to recognize that factors such as the domestic political environment will ensure that their views will never be fully accepted. Instead, they should recognize that there is a good chance that Americans will become involved if there is trouble in the Third World; hence the United States should try to keep order there because it probably won't have the self-restraint to stay out if order breaks down.

Similarly, as realists, the hyper-realists need to acknowledge that a critical component of a country's strength lies in its psychological self-confidence. Even if one can make rational distinctions about what is and is not important to American interests, it matters little if the American people do not accept these distinctions. If the United States is seen to be impotent in the face of challenges from the Third World, even if those challenges do not affect American vital interests, the repercussions of such a stance might well undermine American confidence. This is not to suggest that the United States or any great power should give in to irrational fears. Rather, as meaningless as Third World reverses might be according to some objective criteria, their likely (even if misguided) impact on America's self-image must be taken into account, giving the Third World a significance that a rational analysis might miss.

The Strategic Military Threat Posed by the Third World

The Third World matters both because of the strategic-military threat from the Soviet Union and, more importantly, because of the threat from the Third World states themselves. The hyper-realists underestimate the former and virtually ignore the latter.

The Soviet Threat in the Third World

Despite the momentous changes taking place in and around the USSR, the United States cannot afford to be complacent about future Soviet actions in the Third World. Gorbachev has indeed downgraded the overall importance of the Third World, but he can be toppled or his policies may change. There is already evidence of a renewed Soviet emphasis on some areas of the Third World. Major Third World states such as those of the Association of South East Asian Nations (ASEAN), Mexico, and the Persian Gulf countries are being courted with renewed vigor. Closer ties are being pursued with former enemies such as Israel. Military aid to some radical clients has increased under Gorbachev, and in part because of the need for hard currency, arms transfers are likely to rise in the future. Historically the USSR has consistently escalated the means it employs to gain influence among Third World states—from arms transfers to proxies to direct intervention. The turmoil in the Soviet Union and Eastern Europe notwithstanding, one cannot be confident that this pattern will be reversed.

Should the USSR focus its efforts on extending influence in the Third

World, it has the tools to do so. Despite a weak economy and diminishing ideological appeal, the Soviets are unequalled in getting people into power and keeping them there. Soviet clients are rarely overthrown by coups, insurgencies, or invasions. For would-be and actual Third World leaders this asset is most important in a patron. Third World leaders generally face a far greater number of more serious threats than leaders elsewhere, and loss of power often means execution. Thus they place a high premium on securing a patron who can help defuse the threats to their regimes.

For Third World regimes wanting large numbers of arms, Soviet advantages in speed of delivery (averaging twice as fast as the United States), flexibility in the states to whom they are prepared to sell, large stockpiles, good prices, easy-to-maintain weapons, and availability of proxies have made them the supplier of choice. Consequently the USSR emerged as the leading supplier of arms to the Third World in the 1980s. Compared to the United States, the USSR in the Third World has thirty times more military advisors, trains twice the military personnel, and provides three times the military aid (in dollar terms). To help deal with internal threats, the USSR has placed a security cocoon of advisors from countries friendly to the Soviet Union around Third World leaders (as has been done in Angola, Ethiopia, Mozambique, Zambia, South Yemen, and Libya). As a result, no pro-Soviet regime afforded Moscow's protection has been forcibly replaced by a pro-Western regime since the 1960s, despite the prevalence of coups in the Third World.

Moscow has experienced losses as well as gains in the Third World, but in 1952 the USSR counted only North Korea and China as its allies. Today, countries firmly in the Soviet camp or leaning to the USSR include Cuba, Vietnam, Laos, Cambodia, Nicaragua, Benin, Mozambique, Ethiopia, Yemen, the Congo, Libya, and Syria as well as North Korea. Moreover, Soviet influence cannot be restricted to countries won and lost. The USSR maintains considerable influence in the nonaligned movement, in the United Nations, and with nongovernmental actors such as the Palestine Liberation Organization. Undoubtedly, new setbacks will occur, but it is incontrovertible that the USSR has emerged as a major power in the Third World.

Mirroring its overall record in the Third World, the Soviet Union has had mixed success in securing and maintaining access to military facilities, including bases, among Third World states. Despite several setbacks, the USSR presently makes use of military facilities throughout the Third world, including those in Angola, Ethiopia, South Yemen, Vietnam, Cuba, and most recently in Syria. These facilities assist the USSR in force projection (especially for the navy), provide staging areas for Soviet reconnaissance flights, facilitate intervening in other Third World conflicts, and (especially in Cuba) permit intelligence collection including the monitoring of U.S. communications. The Soviet military presence in these countries may not endanger vital American interests, but it imposes substantial costs. For example, Soviet bases (especially in strategically significant areas such as Central America) hurt American interests by increasing the ability of the USSR to extend its influence and power with fewer forces. The existence of Soviet bases also drives the United States to expend scarce resources to neutralize their impact, thus reducing the American ability to defend other interests.

Most important, access to Third World military facilities matters because they could play a decisive role in the event of a major conventional war between the United States and the Soviet Union. There is a remarkable consensus among Western scholars that the USSR believes that any major conflict with the United States would most likely be conventional, and that Moscow believes it to be in its interest to prevent any escalation to the nuclear level. Similarly, the growing irrelevance of nuclear weapons (except to deter other nuclear weapons) is seen on the United States side as well. Because of changing perceptions, it has become increasingly unlikely that the United States would use nuclear weapons, especially against nonnuclear threats. The lessened credibility of nuclear use, while in many ways welcome, nevertheless increases the possibility of large-scale conventional war between the superpowers.

Preparing for a protracted conventional war dramatically increases the importance of some Third World states by heightening the relevance of traditional concepts of strategy. The likelihood of a major conventional war (at least compared to that of a nuclear war) means that the United States must be concerned about choke points, strategic waterways, land bridges, and sea lines of communication. Consequently, Third World states such as Morocco, Panama, Oman, and the Philippines acquire special importance. Of particular significance is the U.S. ability to secure access to military facilities in time of war while denying them to the USSR. According to one analyst, the threat to sea lines of communication posed by Cuba and—until recently—by Nicaragua could cause the United States to lose a war in Europe.[10] Preparing to neutralize these threats and forestalling additional Soviet footholds consequently becomes a pressing American concern. Moreover, it is critical that the United States be able to project forces into Third World areas that the USSR deems critical for the successful prosecution of a conventional war. According to Michael MccGwire, in a conventional war, the USSR would attempt to maintain a defense perimeter that included Europe, southwest Asia, and northern Africa.[11] By making it clear to the Soviet Union that these objectives would be denied them in the event of war (by, for example, maintaining U.S. access to military facilities in the relevant regions), American deterrence would be enhanced.

Balancing and Bandwagoning

The argument made by Stephen Walt that, because we live in a balancing world, threatening policies by either the United States or the USSR will drive countries to its adversary is dangerously misguided. The argument (which is supported by traditional balance-of-power theory) is an important one because the hyper-realists rely heavily on their analyses of balancing to justify their recommendations that the United States remain essentially aloof from the security concerns of the Third World. After all, if aggressive involvement will simply force countries to the other side, the United States need not actively counter Soviet or other threats in the Third World because the threatened states will flock to the United States on their own.

There are several problems with this argument. Determining that states tend to balance does not tell us much about the prospects for a specific country. If, for example, 90 percent of a given set of countries balanced, that does not mean that there is a 90 percent chance that another country would behave the same way. Thus it is foolhardy to complacently assume that balancing will occur in a given situation simply because it has predominated in another context. The concept of balancing also may not be relevant to many countries in the Third World. As Walt acknowledges, bandwagoning (or appeasing an aggressor) will occur when states are weak and when useful allies are not available.[12] These are important qualifications that seemingly would make many Third World countries susceptible to bandwagoning. Moreover, if states will bandwagon when they cannot count on outside help, an aloof American posture in the Third World might undermine any balancing tendencies.

Most important, the nature of the Third World does not lend itself to simple balancing-versus-bandwagoning formulas in which threatening superpowers drive balancing states into the other camp. Third World leaders typically face a multiplicity of threats. In order to survive in power, Third World leaders will attempt to defeat (or balance against) the most pressing threats. In some cases, this will indeed mean turning against the superpower that backs a threat against it. In many other cases, however, Third World leaders will turn toward (or bandwagon to) the superpower backing a threat against it as the best means of defeating that threat. The Third World leadership does this in the belief that the superpower backing the threat against it is also in the best position to undermine that threat (by, for example, suspending its arms supplies or applying political pressure to its client). Moreover, by bandwagoning to a superpower that backs the threat and not to the country or group that directly poses the threat, the Third World regime does not place its survival in the hands of its greatest enemy. Equally significant, the great majority of threats against Third World leaders are internal (a point virtually ignored by Walt). Therefore Third World leaders seeking outside support to survive in power are at least as likely to align with countries that will protect them from insurgencies, revolutions, and coups as they are to align with countries that will protect them from other threatening states. The concepts of balancing and bandwagoning employed by the hyper-realists ignore this critical domestic dimension and are thus not very helpful in understanding Third World alignment decisions.

Even if Third World leaders choose to balance against American threats by intensifying their alignment with the USSR, the presence of an American-sponsored threat may deprive them of that option by foreclosing Soviet support. After all, Third World states cannot align with the USSR if Moscow refuses to back them. American assistance to groups seeking the overthrow of the pro-Soviet regimes (the Reagan Doctrine) has raised the cost to Moscow of its Third World empire and has helped bring about the alleged downgrading of the Third World in Soviet priorities so ballyhooed by the hyper-realists. In Africa, American and (regrettably) South African aid to the rebel group UNITA helped convince the USSR that its hold on Angola had become increasingly expensive to maintain. By denying the USSR and its allies a military victory, the United States helped induce Moscow to apply pressure to its Cuban and Angolan clients to reach a diplomatic settlement. Most significant, American efforts in Afghanistan have brought about a Soviet withdrawal, which could result in the first-ever success of an insurgency against a Marxist–Leninist government.

None of this is meant to suggest that threatening states is the most effective way for an outside power to secure influence in the Third World. Rather, the complexity of the Third World is such that it is misleading to draw sweeping conclusions about how outside powers should approach the Third World based on abstract notions of balancing and bandwagoning. In attempting to assess how Third World states will react to threats, each situation must be judged on its own merits and special attention must be devoted to the role of internal threats. Because Third World leaders (as leaders anywhere) will turn to the state most likely to keep them in power, policies of benign neglect stand little prospect of attracting many friends.

Third World Threats without Soviet Involvement

Even if the Soviet Union were to remove itself totally from the Third World, the United States would still face major threats to its interests from Third World countries and groups. In their zeal to debunk the Soviet threat and promote American noninvolvement, the hyper-realists have placed little emphasis on these threats. Foremost among them is the specter of nuclear proliferation. At present it is believed that India, Israel, and Pakistan either have nuclear weapons or are very close to developing a nuclear-weapons capability. Several other countries, such as Libya, Argentina, Brazil, and Iraq, are suspected of attempting to buy or develop nuclear weapons. The threat of nuclear proliferation is especially alarming when Third World countries are involved. Unlike in the U.S.–Soviet balance, Third World countries embroiled in intense conflicts might develop vulnerable nuclear forces without sophisticated command and control. The possibility of accidental or deliberate use of nuclear weapons would thus be far greater than has ever existed between the superpowers. Can anyone doubt that the possession of nuclear weapons by either Iran or Iraq (or both) would have led to their use during their conflict? If nuclear weapons are used, the United States or its allies face the possibility of being dragged into a Third World nuclear conflict, perhaps in confrontation with the USSR, or even being the target of a nuclear strike itself.

In addition to the threat posed by proliferation, Third World states are gaining access to other weapons that threaten the security interests of the United States and its allies. At least fifteen Third World countries now possess ballistic missiles; seven maintain active indigenous development programs.[13] The greatest danger posed by ballistic missiles is that they will be combined with nuclear weapons, but it is not the only threat. The speed of ballistic missiles combined with the absence of defenses against them makes these weapons ideal for use in surprise or terror attacks even if nuclear weapons are not used. This is especially the case given the steady improvements in the accuracy and range of Third World ballistic missiles. Moreover, the use of over 500 ballistic missiles in the Iran–Iraq conflict and almost 100 in the recent Gulf war is likely to lessen any taboo against their use in subsequent wars. Although the Western countries have agreed to limit the spread of technology for ballistic missiles, indigenous development and the actions of countries outside the 1987 agreement (e.g., the Soviet Union, China, Argentina, and Brazil) raise the prospects of their likely proliferation throughout the Third World. This, in turn, places American allies (e.g., Israel, Pakistan, and South Korea) at risk and increases the chances that an irresponsible leadership or terrorist element may gain control of these weapons and endanger the United States directly.

Along with ballistic missiles other weapons enhance the threat posed by Third World states to American security interests. At present the Central Intelligence Agency estimates that fourteen Third World states have chemical weapons while an additional ten are trying to make them.[14] Chemical weapons are especially frightening in that small amounts can quickly kill large numbers of people (especially if they are unprotected civilians). Iraq has already used chemical weapons in its war with Iran and against its own Kurdish population. The effectiveness of the Iraqi attacks in blunting Iranian offensives, forcing the evacuation of much of Teheran, and promoting the expulsion of the Kurds, combined with a lack of international condemnation, is likely to make similar attacks by other Third World states more probable in the future. As an indication of American concern over chemical attacks, the United States refused to rule out launching a preemptive strike on a suspected Libyan chemical plant early in 1989. Like nuclear weapons, chemical weapons are especially frightening when carried by ballistic missiles. Syria has reportedly already developed a chemical weapon filled with nerve gas for its *Scud* B missiles, and the Libyans may have the capability to do the same. More ominously, Syria and Iraq (among others) are believed to be developing biological weapons.

Even without exotic weapons, Third World states pose a significant and increasing threat to American security interests. The advent of well-armed, socially mobilized populations enables Third World states to confront the United States without Soviet support. The coming to power in Iran of Ayatollah Ruhollah Khomeini and the Iraqi invasion of Kuwait dramatically hurt U.S. interests, particularly in the endangerment of American allies and access to oil in the Gulf. The growing strength of Third World militaries, some of which (e.g., Iraq and Syria) have military forces larger and better equipped than most European countries, is also cause for deep concern. Their strength endangers American allies (e.g., Israel and the ASEAN countries) and could deter American involvement or retaliation in response to attacks on U.S. interests. Moreover, even the weakest of Third World states can threaten American interests by threatening the lives of U.S. citizens traveling or living in their countries. Just such a threat was used by the Reagan and Bush administrations to justify their respective invasions of Grenada and Panama.

The greatest threat to U.S. bases in the Third World comes not from the USSR, but from indigenous forces (American bases in the Philippines are threatened by a Communist insurgency that receives little support from Moscow). In addition, while terrorism is not yet a major threat to American interests, the destruction of a U.S. Marine barracks in Beirut provides ample testimony to how even primitive technology in the hands of some groups can threaten American forces. Should terrorists gain control of more devastating weapons (e.g., precision-guided munitions or, most horribly, nuclear arms), the threat they pose to U.S. security would increase accordingly.

The Third World also threatens American interests in ways that are not usually thought of in terms of security. As evidenced by a secret presidential directive in 1986, the massive drug trafficking to the United States (most of which originates in the Third World) has become a national security threat. Drugs threaten the security interests of the United States by providing an estimated $300 billion a year from Americans to groups some of which are hostile to the United States, by destabilizing democratic governments friendly to the United States (e.g., Colombia, Peru, Bolivia, and Panama), and by diverting the military from its traditional security mission to drug interdiction. Emigration from the Third World to the United States is also increasingly being seen as cause for concern. From 1977 to 1986, about one million people came to the United States each year. This is three times the annual rate in the years 1925–1965. Most of the recent immigrants are Third World refugees fleeing conflicts in Central America and Asia. Large, uncontrolled influxes of populations have strained the resources of local communities, producing bitterness and occasionally violence. The prospect of internal instability afflicting Mexico will make this problem many times worse. If the United States is to deal with this issue humanely, mitigating the frequency and causes of Third World conflict will be essential.

Finally, the Third World affects American security through its impact on the global environment. There is increasing concern that the burning of tropical forests contributes to the greenhouse effect, in which the climate of the earth is substantially warmed. The burning of forests is a major cause of the greenhouse effect because the fires create a buildup of carbon dioxide that traps solar heat in the atmosphere, warming the earth. It is not yet clear how serious this problem is. Nevertheless, if the warnings of many in the scientific community are accurate, than all countries (including the United States) will be affected. Because virtually all of the tropical forests are in Third World countries, getting the leaders of these countries not to burn their forests can become a pressing interest of the United States.

The Threat to American Economic Interests Posed by the Third World

Many hyper-realists argue that economic factors determine the worth of a country to the United States, and that the Third World is economically insignificant. But the Third World is of substantial and increasing importance, especially when specific countries are considered. The Third World portion of the gross world product grew from 11.1 percent in 1960 to 12.3 percent in 1970 to 14.8 percent in 1980. These figures are dwarfed by the share of the developed states (66.5 percent in 1960, 65.7 percent in 1970, and 62.7 percent in 1980) but nevertheless represent a sizeable and growing share of the world economy.[15] U.S. trade with the Third World has also shown modest growth, with the Third World countries consistently buying over one-third of American exports and providing over one-third of American imports. This was especially the case during the 1970s, when American trade with even the non-OPEC states of the Third World grew faster than American trade with the developed world. As a result, the Third World's share of U.S. imports grew from 27 percent in 1970 to 48 percent in 1980, and the percentage of U.S. exports going to the Third World grew from 31 percent in 1970 to 38 percent in 1980. Although the portion of trade with the Third World has declined slightly in the 1980s (partly because of the lower price of oil), it continued to make up more than a third of American trading activity through the middle of the decade (see Table 1). Similarly, the share of American direct investment in the Third World (which does not count bank loans to Third World countries) has declined slightly since 1970 but still accounts for roughly one-quarter of the United States total.

Even if the economic importance of the Third World as a whole to the United States can be debated, the economic importance of individual Third World countries is indisputable. Of particular significance are the East Asian newly industrializing countries (NICs): Taiwan, Hong Kong, South Korea, and Singapore. Trade with them exploded in the 1980s to the extent that in 1987 U.S. imports from these countries were more than two-thirds as large as those from the

Table 1. Percentage of U.S. Exports Going to the Third World and Percentage of U.S. Imports Coming from the Third World

	Exports (percent)	Imports (percent)
1960	36	41
1970	31	27
1980	38	48
1985	37	35

Source: U.S. Bureau of the Census, Statistical Abstract of the United States: 1973, 94th ed. (Washington, D.C.: USGPO, 1973), Table Number 1292, "Exports and Imports of Merchandise by Continent, Area, and Country: 1960–1972"; Statistical Abstract of the United States: 1987, 107th ed., Table Number 1406, "Exports, Imports and Merchandise Trade Balance by Continents, Areas and Countries, 1975–1985."

European Economic Community (EEC) and Japan. Moreover, their 14 percent share of U.S. imports (in 1987), while smaller than the 20 percent from the EEC, was three times as large as the EEC per dollar of GNP. Taiwan, South Korea, and Hong Kong were in the top ten suppliers to the United States, with Taiwan serving as the fourth largest source of U.S. imports—larger than any European country except West Germany.

Other Third World states are also growing in economic importance. Mexico buys more exports from the United States than any European country (it ranks third in the world behind Canada and Japan) and is a greater source of imports to the United States than any European country except West Germany (it ranks fourth behind Japan, Canada, West Germany, and Taiwan). Other examples (such as Brazil) can also be cited as playing an increasingly important role in the U.S. economy. If present trends continue, by the year 2010 India's GNP will be nearly that of France and Brazil's will be about the same as Great Britain. In economics, as in strategic concerns, it is misleading to draw a distinction between the important states (Western Europe and Japan) and the unimportant states of the Third World.

The economic strength of the Third World, particularly certain states, has a direct bearing on the economic well-being of the United States. As the world's largest debtor nation, the United States must increase its exports of goods and services. As the Mexican example demonstrates, some Third World states have assumed central importance as trading partners for the United States. If these Third World states become unwilling or unable to buy American goods, the U.S. economy will suffer.

Ironically, the economic weakness of some Third World states also makes them a source of concern for the United States. Nowhere is this more apparent than in the issue of Third World debt. From $2.1 billion in 1950, Third World debt mushroomed to $1080 billion in 1987. At the very least, the huge debt impedes the economic growth and threatens the stability (by forcing drastic cuts in domestic consumption) of countries important to the United States, such as Brazil and Mexico. Moreover, the inability of many of the countries ever to repay the debt threatens the viability of Western banks, many of them in the United States. A major default reverberating through the banking system could even seriously damage the world economy.

Some hyper-realists acknowledge the growing economic importance of the Third World to the United States, but they are wrong to suggest that the self-interest of the Third World states will ensure that American economic interests in the Third World will not be threatened. It is a fundamental tenet of realism that security concerns will override economic factors in determining a country's policies. A Third World leadership beset by internal and external threats may not choose or may not be able to choose those policies that maximize wealth when its survival is at stake. The United States cannot rely on its Third World trading partners' continuing to pursue policies that benefit the U.S. if their economic interests are superseded by concerns for basic security. In addition, Third World leaders may be motivated by political or ideological interests to pursue policies that are not in their economic self-interest. That many Third World leaders choose to maintain ruinous Communist systems does not give confidence that economic gain will be their primary goal in dealing with other states. These concerns are especially acute when one considers American dependence on oil and other raw materials.

The United States and its allies remain dependent on foreign oil, much of it from the Persian Gulf. At present, imports make up almost one-half of American consumption, more than 60 percent of West European requirements, and virtually all of Japan's needs. All indications are that use of imported oil will grow for the United States and its Western allies in the next ten years while at the same time the existing excess production capacity (of ten million barrels per day) will decline, largely because of Norwegian and UK North Sea production. It is certain that a major part of the shortfall will have to be made up by the Persian Gulf states, which possess nearly 70 percent of the world's current excess production capacity and two-thirds of the world's known reserves. Consequently, it is virtually ensured that Western dependence on Persian Gulf oil will become greater in the next decade.

Although the United States is not as dependent on Persian Gulf oil as are its allies, an oil cutoff could severely damage American interests. As a member of the International Energy Agency (IEA), the United States is obligated to share oil with its allies in the event of a worldwide shortage. Should such a shortage come about, there would be greater competition for alternative suppliers of oil, increases in the price of oil, and possibly shortages in the United States itself. The creation of the IEA and larger stocks of oil have placed the Western states in a better position to deal with a cutoff than they were in 1973. Nevertheless, the projected greater need for foreign oil in the next decade could still provoke an "every nation for itself" scramble, producing the divisive consequences similar to those that occurred in 1973 within the Western alliance. Even if the IEA and stockpiles prove effective, a major disruption would cause economic hardship for the United States (including inflation and unemployment) and might hurt America's ability to conduct a protracted conventional war.

American allies in Japan and Western Europe are threatened far more by developments in the Persian Gulf than by any direct threat against the countries themselves. There have been sixteen oil supply disruptions since 1950, and possibilities exist for many more. Saudi Arabia (which alone possesses 27 percent of the free world's total of oil), as well as the rest of the Persian Gulf states, faces vulnerabilities that simply do not exist among America's more developed allies. Fundamentalist religious groups, such as the one that seized the Grand Mosque in 1979, could take over the government and sharply curtail oil sales to prevent what they see as a drift into Western decadence. A pro-Soviet group could seize power and then request Soviet assistance and military advisors. Palestinians living in Saudi Arabia, civil war between Saudi clans, military officers unhappy with rising levels of corruption, and dissatisfied Shi'ites are all potential threats to stability. As demonstrated by the Iraqi invasion of Kuwait, the threat of an external attack on Saudi Arabia cannot be discounted. Another oil embargo stemming from another Arab–Israeli war also cannot be discounted, especially as Western dependence on the Gulf increases.

The United States cannot hope to end all of these threats. Nevertheless, American involvement could prove critical in the event of protracted internal instability, external aggression, or even in reversing an anti-American coup. It is not probable that the United States will ever have to play such a role. It is, however, far more likely that should the United States be called upon to defend its interests in Western Europe and Japan, it will be in response to threats in the Persian Gulf rather than in response to a Soviet invasion of its allies.

While not as critical as oil, Western dependence on other raw materials is cause for concern. The United States imports more than 90 percent of its domestic consumption of manganese (needed to manufacture steel), chromium (necessary for jet engine parts), cobalt (needed for high-strength steel alloys), and platinum (used for refining and communication equipment). The United States depends on a handful of states in southern Africa for these minerals. Neither substitutes nor alternative suppliers are readily available.

Although steps such as stockpiling would mitigate the damage done to the United States, it is clear that a cutoff could seriously hurt the United States and its allies. The immediate impact would be higher inflation and economic disruptions resulting from production cutbacks among the Organization for Economic Cooperation and Development (OECD) states. In the medium term (five to ten years) failure to adapt to the new technologies could result in major dislocations in the production of stainless and other types of steel. Over the long term, there is no alternative to reliance on some of the minerals, such as chromium.

Far more likely than a decision to launch an embargo is the possibility that widespread instability will prevent the export of strategic minerals. Zaire and Zimbabwe, for example, are major producers of chromium, cobalt, and manganese. Both countries have experienced severe internal strife and regional conflict. Given the staggering array of threats faced by the southern African states and their dependence on primitive transportation facilities to export their products, events in the region may bring about a de facto cutoff that would seriously damage American interests. It is this prospect of chaos rather than Soviet or other hostile control of strategic minerals that presents the greatest threat to American interests. Some of the hyper-realists acknowledge this possibility but, in their zeal to debunk the Soviet threat and emphasize the logic of the marketplace, do not afford it much attention.

It is beyond the scope of this work to detail how the United States should deal with the possibility of chaos threatening the supply of strategic minerals. Under the best of circumstances American leverage will be limited. Nevertheless, it is clear that relying on the laws of economic rationality will not in itself be sufficient. Only an American policy that is engaged in helping bring order and stability to southern Africa and other regions can be relied upon to mitigate this threat.

Just as the hyper-realists exaggerate the role of economic factors in determining a Third World country's interests, they exaggerate the role of economic factors when determining American interests. This focus on economic factors is convenient for the hyper-realists because they are largely correct that most Third World countries, regardless of ideology, will trade with the United States. Consequently they can argue that no Third World countries, even radical, pro-Soviet ones, threaten the United States. This view ignores Third World countries that can be supportive of the U.S. economy and yet threaten some more important set of interests. That Angola is good for Gulf Oil does not change the fact that for ideological and strategic reasons the United States may not wish to see a Marxist–Leninist government in control of an important southern African country. Similarly, just because American companies have profitable investments in Libya and South Africa does not mean that the United States does not have an interest in seeing the regimes of those countries undergo fundamental change. Precisely because big business does not control American policy, and interests other than economic are relevant to American behavior in the Third World, countries can be excellent trading partners and still be hostile to broader American interests.

The Threat to American Political–Ideological Interests from the Third World

The Third World matters because the effort to spread American values matters. The hyper-realists tend to denigrate the importance of ideology; because they believe ideology should not matter, they argue it does not matter. One cannot calculate the value of a country to the United States in terms of the character of its government in the same way that one can calculate its value by virtue of its GNP or military forces. Nevertheless, promoting American values in the Third World is and should remain a critical component of U.S. foreign policy.

It is important for the United States to promote freedom and democracy in the Third World because Americans believe this to be the best way of life. The Third World encompasses most of the world's people. With Western Europe already democratic and essentially secure, and Eastern Europe rapidly becoming democratic, the Third World stands out as one of the last arenas where the extension of American values can take place. If attempting to spread freedom and democracy is good in its own right, ignoring how most of the world's people are ruled simply because they live in countries lacking some objective measure of power is simply wrong.

Extending American values to the Third World is also important in that it can enhance other American interests. None of the democractic countries (with the exception of Finland and India) are aligned with the Soviet Union. The fact that democratic states rarely if ever go to war with one another supports the American interest in stability (at least as far as friendly countries are concerned). Democracies also make better allies. Because democratic governments usually do not assert control over other states, they avoid the disputes common to alliances with transnational ideologies such as Pan-Arabism or Communism. The spread of democracy in the Third World is also important because it can weaken the USSR. The Third World is the last remaining area where Marxist–Leninist ideology is still afforded respect. The lack of receptivity to Marxism–Leninism in the Third World would clearly signal the obsolescence of Soviet ideology, diminishing any contribution it makes to the power of the USSR.

The promotion of American values in the Third World also matters because such promotion is likely to persist. As Robert Tucker argues, all great powers seek to spread their ideology to other states to make certain that their way of life and institutions are not isolated from the rest of the world.[16] There is no reason to believe that the United States will be an exception. Moreover, as even some of the hyper-realists acknowledge, the American people have never been driven by a narrow conception of the national interest or realpolitik. Instead, they are and will continue to be concerned with the nature of regimes in the Third World and ways of bringing them more into conformity with U.S. values.

A TRULY REALISTIC APPROACH TO THE THIRD WORLD

American interests threatened by developments in the Third World are too important and too numerous to ignore. Instead of abandoning the Third World, the United States needs a modulated policy that recognizes the limits of our power but also does not shrink from involvement where we can do some good. Preparations to protect our interests from Third World threats need not be massive or expensive. By employing prudent policies that are sensitive to Third World conditions, the United States can safeguard its interests from actual and potential Third World threats at an acceptable cost and risk.

A successful American policy in the Third World first needs to focus on the security concerns of Third World leaders. These leaders typically face a multiplicity of internal and external threats that endanger their hold on power and their personal survival. Their turning to the outside state that is most likely to ensure their hold on power is as prudent as it is predictable. The hyper-realists are right to point out the increasingly important role played by economics in the affairs of Third World states, and economic aid and trade should remain significant American instruments of influence. But economic interests will not predominate in an environment of insecurity. As seen by the pro-Soviet alignments of Angola and Ethiopia, the potential benefits of long-term economic gains will mean little to Third World leaders facing short-term security threats. There is no substitute for employing military instruments such as arms transfers and advisors in order to meet the security needs of Third World leaders. The use of these instruments by the United States, whether to defend or to threaten, must aim to persuade the leaders of Third World states that their security needs are best met by policies that do not jeopardize American interests.

U.S. policies will increasingly have to deal with Third World challenges that have little or no connection to the USSR. Prominent among these is how to deal with Third World revolutions. The United States should not rush to accommodate apparent Third World revolutions. Before a movement seeks power, the United States needs to try to determine whether it truly represents a mass-based movement for social change or if it is a narrowly supported insurgency. If the former, there is little the United States can do at an acceptable cost. If the latter, and if the regime emerging from the upheaval is likely to pursue policies against American interests, assisting groups attempting to repress the insurgency may very well be the correct policy. The at least initial success of containing the El Salvador insurgency demonstrates the fallacy of prematurely assuming that a guerrilla movement represents an inevitable revolution.

Nor should the United States rush to appease hostile groups that have assumed power. If these regimes attack American interests and violate American values, it makes little sense to support them. Economic aid and trade might be extended if there are prospects that a hostile regime will moderate its behavior.

Moreover, the United States needs to make sure that it is not pushing regimes into the hands of the Soviets or creating hostility where none exists. But once it is determined that the regime (be it on the right or left) is implacably opposed to American interests regardless of U.S. attempts at conciliation (as was the case in Ethiopia, for example), Washington need not apologize for opposing it. Not only would appeasement not work, but it must be remembered that there is much to lose in accommodating regimes whose practices and goals are repugnant to the American people. The American experience with Panama under General Manuel Noriega is a case in point.

In the Third World, as elsewhere, the United States must set priorities. The hyper-realists are certainly correct in arguing that, given limited resources, the United States cannot pursue all interests with equal vigor. Moreover, the United States must avoid the danger of spending scarce capital (both human and financial) on peripheral interests. The hyper-realists' fear of another Vietnam, however, has prevented them from realizing that an activist, involved U.S. policy safeguarding U.S. interests in the Third World can be pursued at a reasonable cost and at no threat to a "peace dividend" should one materialize.

Such a policy would begin with the assumption that no interest in the Third World is worth another protracted, Vietnam-like war. Instead, the United States must be prepared to protect vital interests, such as in the Persian Gulf, with the ability to launch a major intervention that can accomplish its goals quickly. On the other end of the spectrum, the United States must also be able to mount low-level operations and provide economic and military assistance so that the vast majority of its other interests in the Third World can be safeguarded as well. This is a formula not for a resurgent America, but for an engaged one.

On the need to be able to intervene in the Persian Gulf, there is little dispute. There is some disagreement as to just what is needed, but the hyper-realists accept the premise that such a capability is necessary. There is no reason why the interventionary capability developed primarily for the Persian Gulf could not be used for other Third World contingencies as well. The United States is thus left with the requirement of maintaining Third World interventionary forces whether one believes they are needed outside the Gulf or not.

The specific kinds of forces that the United States will need to be prepared to intervene go beyond the scope of this study. Nevertheless, as the Soviet threat to Europe diminishes and the actual and relative threat from the Third World to American interests increases, some changes in force structure will be needed. Instead of placing so much emphasis on heavy armor and large numbers of troops stationed in Europe, the United States will have to redirect some of its efforts towards Third World challenges. This will involve a greater effort to develop highly mobile, lightly armed forces that can reach the site of a conflict quickly. Because many Third World forces are so well armed, the United States must also be prepared to intervene with heavy armor where necessary. Making sure that

American forces are heavy enough to fight in the Third World, but not so heavy that they cannot be quickly transported, is a challenge warranting serious preparation in the coming years.

While the hyper-realists accept the need of the United States to intervene in the Third World for vital interests, they are sharply critical about American efforts to protect its less-than-vital Third World interests. This is somewhat curious as the United States can protect its less-than-vital Third World interests at a relatively low cost. Although numerous and potentially serious, the vast majority of threats to American interests in the Third World can be defeated without direct American intervention or the expenditure of large sums. As the *Report of the Commission on Integrated Long-Term Strategy* states, a program to deal with low-intensity Third World threats that included increased security assistance for friendly states and support for anti-Communist insurgencies would cost only 4 percent of the U.S. defense budget, resulting in expenditures of $12 billion per year.[17] Even for nonvital Third World interests, this sum can hardly be considered excessive, nor would it leave the United States unable to confront more important challenges. If Washington is serious about reducing its defense budget, it must look not to the Third World, but to Europe, where the bulk of the defense dollar is spent.

In the final analysis the preceding debate is less about costs or even tactics than it is about the attitude one should adopt toward the Third World. It is indisputable that the United States needs to give priority to protecting itself and its allies in Western Europe and Japan. But the United States also needs to recognize that, in large part because of American efforts, the prospect of a Soviet invasion of Europe or a nuclear war between the superpowers has become extremely remote. In the meantime, the great majority of the world's states and people are marked by new levels of power and instability. To ignore the threat they pose and will continue to pose to American interests would be the most unrealistic policy of all.

NOTES AND REFERENCES

1. An earlier version of this chapter appeared in *International Security* 14 (Summer 1989), pp. 50–85.
2. I have defined the Third World in general conformity with the categorization used by the United Nations to include all countries except the United States, Canada, Japan, Australia, New Zealand, South Africa, the European states, and the Communist states of Asia.
3. The school of thought that this article addresses is derived principally from the following pieces of work: Robert H. Johnson, "Exaggerating America's Stakes in Third World Conflicts," *International Security* 10 (Winter 1985–1986), pp. 326–68; Richard E. Feinberg and Kenneth A. Oye, "After the Fall: U.S. Policy toward Radical Regimes," *World Policy Journal* 1 (Fall 1983), pp. 201–15; Jerome Slater, "Dominos in Central America: Will They Fall? Does It Matter?" *International Security* 12 (Fall 1987), pp. 105–34; Barry Posen and Stephen Van Evera, "Reagan

Administration Defense Policy: Departure from Containment," in Kenneth A. Oye, Robert J. Lieber, Donald Rothchild, eds., *Eagle Resurgent? The Reagan Era in American Foreign Policy* (Boston: Little, Brown, 1987), pp. 75–114; Stephen M. Walt, *The Origins of Alliances* (Ithaca: Cornell University Press, 1987); and Richard E. Feinberg, *The Intemperate Zone: The Third World Challenge to U.S. Foreign Policy* (New York: Norton, 1983). For a concise statement of the hyper-realist position, see Stephen Van Evera, "American Strategic Interests: Why Europe Matters, Why the Third World Doesn't," testimony prepared for hearings before the Panel on Defense Burdensharing, Committee on Armed Services, U.S. House of Representatives, March 2, 1988.

4. The term *hyper-realists* was chosen because of the strong adherence of these authors to the realist school of international politics, especially as practiced by George Kennan, Walter Lippmann, and, to a lesser extent, Hans Morgenthau. The *hyper* prefix stems from their taking the emphasis of Kennan and others on material, objective factors (as compared to other factors such as ideology) to its illogical extreme. I am indebted to Aaron Friedberg for the use of the term *hyper-realists* to describe this school of thought.
5. Posen and Van Evera, "Reagan Administration Defense Policy," p. 97.
6. Quote is from Slater, "Dominos in Central America," p. 124.
7. Feinberg, *The Intemperate Zone*, p. 109.
8. Feinberg, *The Intemperate Zone*, p. 240.
9. Walt, *The Origins of Alliances*; see especially p. 282.
10. Michael C. Desch, "Turning the Caribbean Flank: Sea Lane Vulnerability during a European War," *Survival* 29 (November–December 1987), p. 524; Desch believes that Soviet use of Nicaragua to interdict American shipping would double the threat posed by Cuba.
11. Michael MccGwire, *Military Objectives in Soviet Foreign Policy* (Washington, D.C.: The Brookings Institution, 1987), pp. 48–49, 213–31.
12. Walt, *The Origins of Alliances*, pp. 173, 175.
13. See lists by Aaron Karp, "Ballistic Missiles in the Third World," *International Security* 9 (Winter 1984–1985), p. 176; W. Seth Carus, "Missiles in the Middle East: A New Threat to Stability," *Policy Focus*, Washington Institute for Near East Policy, no. 6, June 1988; and *Supporting U.S. Strategy for Third World Conflict* (Washington, D.C.: U.S. Department of Defense, June 1988), p. 13.
14. *Newsweek*, September 19, 1988, p. 30. The list includes Iraq, which is confirmed to have chemical weapons, and Egypt, Syria, Libya, Israel, Ethiopia, Burma, Thailand, North Korea, Cuba, Iran, Vietnam, Taiwan, and South Korea, which are believed to have them. The People's Republic of China and South Africa (not usually categorized as Third World states) are also included on the list of countries suspected of having chemical weapons.
15. Percentages are based on figures from the U.S. Central Intelligence Agency, National Foreign Assessment Center, *Handbook of Economic Statistics, 1981* (Washington, D.C.: USGPO, 1981), Table 9. Percentages calculated by Kenneth A. Oye and can be found in Kenneth A. Oye, "Constrained Confidence and the Evolution of Reagan Foreign Policy," in Kenneth A. Oye et al., *Eagle Resurgent?*, p. 10.
16. Tucker does not, however, believe that it is necessary for the United States to spread its values to the Third World. See, for example, "The Purposes of American Power," *Foreign Affairs* 59 (Winter 1980–1981), pp. 241–74.
17. Report of the Commission on Integrated Long-Term Strategy, *Discriminate Deterrence* (Washington, D.C.: USGPO, January 1988), p. 16.

10

New Weapons and Old Enmities
Proliferation, Regional Conflict, and Implications for U.S. Strategy in the 1990s

Lewis A. Dunn

Since the late 1980s a remarkable and accelerating process of political and military change has dominated relations between East and West. Under the leadership of Mikhail Gorbachev, *glasnost*, *perestroika*, and Soviet new thinking have opened up the possibility of a fundamental transformation of the U.S.–Soviet relationship. Arms-control agreements are near completion that will begin a process of building down both sides' nuclear-, conventional-, and chemical-weapons capabilities. With Soviet tolerance, if not encouragement, Communist political monopoly and social control were overthrown in Poland, Hungary, Czechoslovakia, East Germany, Bulgaria, and Rumania. These political changes, symbolized by the opening of the Berlin Wall in 1989, resulted in the first steps to overcome Europe's four decades of political, economic, and social division. Politicians on both sides declared that the Cold War had ended.

Recent developments and future prospects in the Third World stand in stark contrast to this growing breakdown of East–West confrontation. On the one hand, long-standing political, military, and ethnic enmities remain the distin-

---

Lewis A. Dunn • Science Applications International Corporation, McLean, Virginia 22102.

guishing feature of regional politics in the Middle East, South Asia, the Persian Gulf, and elsewhere. On the other hand, the proliferation of advanced weaponry—top-of-the-line conventional weapons, ballistic missiles, chemical weapons, and nuclear weapons—stands out as the most prominent aspect of change in virtually all of the globe's regions.

The proliferation of advanced weaponry will increase the likelihood, scope, and intensity of Third World conflicts. It will also pose new challenges for U.S. policymakers. Not least, it will require a new way of thinking about the problem of proliferation, one which recognizes that traditional efforts to contain the spread of advanced weaponry will need to be joined to new policies to contain the regional consequences and global spillovers of proliferation.

THE PROLIFERATION OF ADVANCED WEAPONRY

The proliferation of advanced weaponry in conflict-prone regions throughout the Third World will continue and accelerate in the 1990s. Consider both the overall pattern and some of the more striking region-by-region developments.

From Advanced Conventional Weapons to Weapons of Mass Destruction

Sales of conventional arms to Third World countries are not a new phenomenon. The United States, the Soviet Union, Eastern and Western European countries, Brazil, and China have supplied hundreds of billions of dollars in arms to friends, allies, and willing purchasers over recent decades. In some respects, however, the types of conventional weapons being sold and the conditions under which they are sold are changing. Traditional suppliers of conventional weapons have become more ready in recent years to supply first-line conventional equipment to their Third World purchasers than they have been in the past. American sales of F-16 fighter aircraft to Pakistan, Soviet sales of the Su-24 fighter-bomber to Libya, and British sales of the multirole *Tornado* aircraft to Saudi Arabia exemplify this trend. Of even greater importance, the advanced conventional weaponry now being sold to many Third World countries has enhanced capabilities for power projection. Longer range aircraft, aerial refueling capabilities, and airborne warning and control systems provide an ability to use force and to intervene in conflicts at great distances. Not least, sales of advanced weaponry have increasingly been accompanied by transfers of technology to build up these countries' own domestic arms-manufacturing capabilities.

The growing proliferation of ballistic missiles—along with the diffusion of the technologies and the manufacturing capabilities to make them—is part of the change now under way in Third World arsenals. For now, short-range missiles (about 300 kilometers) predominate. But extended range systems (from 900 to

1,500 kilometers) are becoming available, usually through semi-indigenous product improvement of an earlier shorter range system. Still other longer range systems (several thousands of kilometers) could be widely sold or developed in the future.

Moreover, the number of countries with programs to develop ballistic missiles, either independently or in cooperation with other developing countries, has steadily grown. Director of Central Intelligence William Webster estimated that fifteen Third World countries will produce ballistic missiles by 2000.[1] Nonetheless, some important thresholds for ballistic-missile proliferation in the Third World still exist. The numbers of missiles and launchers vary considerably but frequently are in the tens, not the hundreds. With some exceptions the range of the ballistic missiles now held by most Third World countries still is short, and the accuracy, reliability, and sophistication of most of these missiles remain limited.

Chemical-weapons proliferation is also accelerating in the Third World. In the last decade the number of countries with or pursuing such a capability is estimated to have jumped to nearly two dozen.[2] Successful Iraqi use of mustard, cyanide, and nerve agents during the 1980s Gulf War with Iran—as well as the lack of global condemnation of Iraq's actions—contributed to this jump. And although evidence is still very scant, there is growing concern that some countries are beginning to think seriously about the acquisition of biological weapons.

Though overshadowed in the past several years by fears of missile and chemical-weapons proliferation, the development of nuclear weapons by additional countries in the Third World is increasingly likely in the 1990s. There are eight to ten countries that are considered either to be seeking a nuclear-weapons capability or are likely to do so if a regional rival successfully takes that route. Initially such proliferation probably will be unacknowledged, following what is widely believed publicly to be the Israeli model. But overt proliferation may also occur. Nuclear-weapons proliferation in the 1990s can be expected to go beyond the production of a few simple, last-resort fission nuclear weapons. Instead, some new nuclear-weapon states will begin advanced nuclear-weapons proliferation, adding to the numbers and sophistication of their nuclear warheads and delivery vehicles, developing explicit nuclear strategy and doctrine, and generally taking decisions associated with the creation of a nuclear force.

There is under way an overall process of proliferation of advanced weaponry in Third World regions with many linkages among its separate elements. New power-projection capabilities in one country, for example, have added to incentives to acquire chemical or nuclear weapons in others, even though these capabilities may make it easier for a country already possessing such weapons to block their acquisition by its neighbors. In addition, ballistic missiles have magnified the potential threat of either chemical or nuclear weapons by offering what is perceived to be a potentially unstoppable delivery system.

Regional Proliferation Trends

Evidence of the proliferation of advanced weaponry stands out on a region-by-region basis. A region-by-region survey makes clear the need to focus attention not exclusively on chemical and missile proliferation, as has been the case recently, but also on nuclear-weapons activities of Third World countries.

The Middle East

The Middle East (from Libya in the west to the Persian Gulf and Iran in the east) exemplifies the preceding elements. Top-of-the-line conventional weapons, ballistic missiles, and chemical and nuclear weapons figure prominently in any sketch of Middle East military capabilities. The conventional military balance of an earlier day is giving way to complex and overlapping networks of military balances—conventional versus conventional, chemical versus conventional, chemical versus chemical, chemical versus nuclear, nuclear versus chemical and conventional, and before too long nuclear versus nuclear.

Over the past decade, countries in the Middle East have been the main purchasers of conventional arms. For the most part, however, Israel alone has had a significant capability to project military power beyond its borders, as illustrated both by its 1976 rescue mission to Entebbe and its destruction of Iraq's Osirak nuclear research reactor in 1981. In the years ahead, Israel's Arab neighbors will add to their power-projection capabilities. Purchases of longer range aircraft, for example, will give Libya, Saudi Arabia, and perhaps Iraq—if restrictions on its future arms purchases prove ineffective—new capabilities to intervene in a future Arab–Israeli conflict.

Ballistic-missile capabilities have become a prominent part of the military arsenals of countries in this region. Egypt, Iran, Iraq, Libya, North Yemen, South Yemen, and Syria possess short-range ballistic missiles (e.g., *FROG*-7s, *Scud*-Bs, or SS-21s) supplied either by the Soviet Union or by second-tier suppliers that had received such missiles from the Soviet Union. Saudi Arabia's 1988 purchase of Chinese-made CSS-2 missiles with a range of 2,500 kilometers brought a new dimension to regional missile proliferation, although these have yet to become operational. For its part, Israel is reported to have deployed both the *Jericho* I and the *Jericho* II missiles, while testing a *Jericho* IIB with an estimated range of 1,400 kilometers.

In the decade ahead it must be expected that the ballistic-missile capabilities of Middle East countries will continue to grow in numbers, range, and sophistication. During its war with Iran, Iraq (possibly with North Korean help) developed a 600-kilometer version of the *Scud*-B, permitting it to attack Teheran. Before the recent war, work was under way on a longer range (2,000-kilometer) system. Libya, apparently, is working on an extended range version of the *Scud*-B, which would give it a capability to strike targets in Israel as well as U.S.

naval bases in the Mediterranean. In turn, Israel's successful launch of a space satellite in 1988 demonstrated its capability to produce longer range ballistic missiles able to strike distant targets, including Moscow.

The spread of chemical-weapons capabilities to countries in the Middle East will compound the impact of the growth of power-projection capabilities as well as the acquisition of ballistic missiles. Iraq and, to a lesser degree, Iran used chemical weapons during the Iran–Iraq conflict, and both countries have been trying to enhance their capabilities. Egypt, Libya, and Syria as well as Israel are also thought to have chemical-weapons capabilities or programs under way.

By contrast it is widely assumed that the status of nuclear proliferation in this region is relatively stable: Israel will not shift from its posture of unacknowledged but widely assumed possession of nuclear weapons, and its Arab enemies and Iran either will be too technically backward to acquire nuclear weapons or will be prevented from doing so by Israeli preventive military action. Growing pressures, however, could result in the erosion of the Middle East nuclear status quo in the 1990s. Prewar public reports, for example, suggest that Iraq was again seeking to acquire a nuclear-weapons capability.[3] Iraq has already demonstrated its technical, organizational, and engineering capabilities in the missile and chemical fields. Besides, Iraq is publicly rumored to have access to outside assistance.

Until the 1991 war, it would have been far harder for Israel to locate and then destroy Iraq's nuclear-weapons infrastructure in a repeat of its 1981 raid because of dispersion of Iraq's activities, new active and passive defensive measures, and use of underground facilities. Iraq demonstrated its ability to fire ballistic missiles at Israel, including one apparently aimed at the Dimona nuclear complex.

Unless the conditions imposed by the U.N. on Iraq for a permanent cease-fire to the 1991 Gulf war are effective indefinitely, a renewed Iraqi nuclear-weapon program could not but increase motivations in Iran, Saudi Arabia, and Syria to acquire nuclear weapons. While technical constraints would be a major obstacle to Saudi or Syrian nuclear-weapon programs, Iran probably would be able to match an Iraqi bomb, though with a delay of some years. Iraqi acquisition of nuclear weapons would also weaken support for nuclear restraint in Egypt, a country that thought seriously about acquiring nuclear weapons in the early 1970s. Iraqi success would certainly result in a major expansion of the Israeli program, with public deployments to buttress deterrence and possibly nuclear testing to move to more advanced nuclear warheads.

South Asia

India and Pakistan will increase their conventional power-projection capabilities in the decade ahead with planned deployments of longer range aircraft, aerial refueling, and aerial-surveillance capabilities purchased from their respective Soviet and American allies. Both countries will continue to develop ballistic missiles, though India would retain its considerable lead should it go ahead with the fielding of its 2,500-kilometer-range *AGNI* missile (compared to Pakistan's slower project to develop 300- and 600-kilometer missiles) and the possible development of an intercontinental ballistic missile (ICBM). Although the evidence is more uncertain than that for ballistic-missile development, India and Pakistan may be taking initial steps toward a chemical-weapons capability.[4]

Both countries already have considerable nuclear-weapons-related expertise. India detonated a nuclear device in May 1974 but then essentially put its nuclear-weapon program on hold. Spurred on by this Indian nuclear test, Pakistan mounted an intensive effort to acquire nuclear-weapons material, master the technology of nuclear weapons, and lay the foundation for a nuclear-weapons capability. By the late 1980s, public statements by Pakistani leaders made clear that Pakistan possessed the capability to make nuclear weapons should a decision be taken to do so. Indian officials made comparable public statements, as well as warning of the possible need to reassess their own nuclear restraint. Nonetheless, both countries' leaders appear to have been reluctant to move across the nuclear-weapon threshold.

As in the Middle East a mix of pressures could break down this relative nuclear restraint. In either country, bureaucratic and scientific momentum could eventually carry one or the other country to the unacknowledged production of nuclear weapons. At some future time, moreover, Pakistani nuclear creep may make it impossible for the U.S. president to certify to Congress that Pakistan does not possess a nuclear device (as required by law if American economic and military aid is to continue). This would make it very difficult for the Indian government to resist public and elite pressures for reinvigorating India's nuclear-weapon program. Increased political and military tensions between India and Pakistan, for example, over the future status of Kashmir or allegations of Pakistani support for separatists in India could provide a different incentive for either country to deploy nuclear weapons. Pakistani deployment of extended range ballistic missiles capable of striking major Indian cities also could trigger Indian nuclear-weapons activities. Beyond the region, acquisition of such weapons by Iraq or Iran also would strengthen the case of those in India and Pakistan who have been arguing for nuclear deployments.

The manufacturing and stockpiling of nuclear weapons by Pakistan and India initially could be undertaken covertly, although rumors would occur. Such a period of limited and unacknowledged proliferation, however, is likely to be unstable. Efforts by India to trump Pakistan's nuclear capability by developing thermonuclear weapons, worst-case military analysis in India, and affronted Indian domestic pride are but three reasons why, if India's leadership decided to resume that country's nuclear-weapon program, it probably would not be content with a modest, unacknowledged bomb-in-the-basement capability. For its part Pakistan could also seek an open capability in order to present the United States with a fait accompli and to call for restored U.S. aid on the basis of the new realities.

The Korean Peninsula and Taiwan

The Korean Peninsula stands out as another potential proliferation hot spot. Conventional forces on both sides of the demilitarized zone will grow more sophisticated, as exemplified by projected deployments of U.S. F-18 and Soviet MiG-29 aircraft. North Korea already is publicly reported likely to possess chemical weapons, while South Korea may be seeking a comparable capability. Even more a cause for concern is the possible deployment of ballistic missiles and nuclear weapons by both countries during the 1990s.

In particular, North Korea has become both a producer and an exporter of a modified version of the Soviet *Scud*-B missile. Efforts are likely to extend the range of that system to more than 600 kilometers, giving it a capability to target all of South Korea. Depending on future Chinese policies, North Korea could also be a purchaser of China's M-9 missile with an estimated range of 600 kilometers. South Korea has modified U.S.-supplied *Nike* surface-to-air missiles for a surface-to-surface mission and has produced its own surface-to-surface missile. Looking further ahead, South Korea is beginning a space launch program, which could well be the source of longer range missiles by the mid to late 1990s.

The possibility of nuclear-weapons proliferation on the Korean Peninsula is also increasing. North Korea has completed a large research reactor that will be capable of producing plutonium to make nuclear weapons. Public reports also suggest that it may be building a plant to reprocess the spent reactor fuel to obtain plutonium.[5] Although North Korea has acceded to the Treaty on the Non-proliferation of Nuclear Weapons (NPT), as has South Korea, Pyongyang has been dragging its feet in concluding a required agreement to permit international inspection of its peaceful nuclear activities.

North Korea's nuclear activities have generated considerable uneasiness and suspicion in South Korea. Warnings about the dangers of a North Korean nuclear-weapons capability have become more frequent than in the past. Should North Korea continue to resist international inspection—let alone acquire a nuclear-weapons capability—there is every reason to believe that South Korea would revive the publicly reported nuclear-weapons-related activities that it had under way in the mid 1970s and later stopped under pressure from the United States. Major reductions of U.S. ground and air forces in South Korea, which appear increasingly likely to occur, would reinforce motivations to acquire nuclear weapons. Such reductions would also eliminate a major source of U.S. leverage over Seoul's nuclear ambitions. There is little doubt, moreover, that with its industrial and technical base South Korea could rapidly produce nuclear weapons after a decision to do so.

Though its conventional forces remain limited, Taiwan is thought to have chemical weapons and is developing an extended range (1,000-kilometer) missile, perhaps with Israeli assistance. Equally important, Taiwan had a nuclear-weapon program under way in the mid 1970s, only to be forced by the United States to put it on the shelf. Recent public disclosures of Taiwan's renewed attempts to obtain separated plutonium suggest that its nuclear ambitions may remain close to the surface.[6]

Southern Africa

The proliferation situation in southern Africa is mixed. Although South Africa's access to advanced conventional weaponry is limited, it apparently is about to produce a ballistic missile with Israeli assistance. Similarly, while official South African statements have made clear that it could produce nuclear weapons (should it so decide), motivations to do so may decline in the 1990s.

South Africa's nuclear-weapons-related activities apparently stemmed partly from South African worst-case analyses in the 1970s about the risks of outside military intervention to overthrow the white apartheid regime. A capability to make nuclear weapons also provided a potential means to bolster white morale to hold out against foreign political and economic sanctions. Recent political changes in South Africa, however, suggest that the government may be prepared over a period of years to work out an accommodation with eventual black rule. The settlement in Angola and the withdrawal of Cuban troops have also lessened fears of outside intervention. Many of the motivations for nuclear weapons, therefore, may no longer hold. Indeed, there are indications that South Africa will decide to join the NPT in an attempt to gain some international legitimacy, thereby putting its nuclear program under international supervision, as well as taking steps to reassure other countries about the extent of its activities prior to joining.

The Southern Cone

In South America the proliferation situation is less threatening. Budgetary constraints have hampered the modernization of conventional military forces in both Brazil and Argentina. At the same time, Brazil and, to a lesser degree, Argentina have emerged as exporters of conventional arms (e.g., aircraft trainers, small arms, and armored fighting vehicles).

Consistent with that export approach, Brazil has a wide array of programs

grams under way to develop and export ballistic missiles, sounding rockets, and space launch vehicles with ranges from several hundred to over one thousand kilometers. Argentina had also been seeking to develop a ballistic missile in cooperation with Iraq (and previously Egypt), but that program now appears to be on hold. Both countries, however, could develop ballistic missiles but then export them rather than deploy them. What course they choose will depend on the limits of their technical capability (neither country has yet deployed even short-range systems) and, even more, will be dependent on events in the nuclear arena.

Under former Presidents Raul Alfonsin and Jose Sarney, Brazil and Argentina began a dialogue on nuclear issues, with exchanges of visits to sensitive nuclear facilities in each country. As a result of these steps, the transparency of each side's nuclear program has increased and suspicions about the nuclear-weapons ambitions of the other have been reduced. Although both Sarney and Alfonsin have left office, this dialogue has continued. Motivations to move closer to a nuclear-weapons capability out of fear that the other side is doing so have been reduced in this region. (Brazil has publicly ended its nuclear-weapon program.)

Nonetheless events outside the region still could lead to a jump in each country's incentives to acquire nuclear weapons in the decade ahead. In particular, should India, Iraq, or Pakistan acquire nuclear weapons, Brazilian leaders might choose to renew their program on the grounds that national prestige and Brazil's role in world affairs demand no less. If so, an Argentine nuclear-weapon program would result, as would deployment of ballistic missiles by both countries. The current improving political relationship between these two countries could still give way to a growing military rivalry.

ADVANCED WEAPONRY, REGIONAL CONFLICT, AND GLOBAL SPILLOVERS

The increasing proliferation of advanced weaponry has important implications for the patterns and characteristics of regional conflict in the 1990s. New threats and difficult choices will also confront outside powers, especially the United States and the Soviet Union.

Patterns of Regional Conflict

New weaponry will heighten the complexity of existing military confrontations in the Third World, with the costs and benefits of using military force difficult to calculate. Crisis instabilities could be particularly high, and any transition to stable deterrent relationships between traditional rivals prolonged, difficult, and not ensured. Because of these instabilities, the likelihood of conflict could increase significantly. In addition, missile, chemical, and nuclear proliferation will expand the scope and increase the intensity of regional conflict.

New Patterns of Military Confrontation and Deterrence

The proliferation of advanced weaponry, especially ballistic missiles armed with chemical or nuclear warheads, will increase the complexity of military confrontations in the Middle East, South Asia, the Persian Gulf, and Northeast Asia. In some instances regional rivals such as Iraq and Iran will possess chemical-weapons capabilities. In other cases one regional rival may possess chemical weapons (Syria or Iraq) and another may have nuclear weapons (Israel). Other possibilities include confrontations in which both sides possess nuclear and chemical weapons (e.g., between Iraq and Israel in the future) or both possess nuclear weapons but not chemical weapons (e.g., perhaps Pakistan and India).

The relative sophistication of individual countries' capabilities will vary as well. With regard to nuclear proliferation, differences can be anticipated, for example, in the numbers and types of nuclear warheads (from a few crude fission weapons to over a hundred advanced nuclear weapons), delivery systems (from aircraft to vulnerable missiles to survivable missiles), physical security and command and control, and organization. Chemical-weapons capabilities are likely to vary in terms of types of weapons; stockpile durability, usability, and reliability; and means of delivery.

These differences could be quite striking in some regions. It is highly unlikely, for instance, that a successful Iraqi nuclear-weapon program would catch up with Israel's far more diversified and advanced nuclear-weapons activities. Similarly, in a future nuclear arms race, India would be likely to outpace Pakistan in numbers of warheads as well as numbers, range, and reliability of missiles for delivery. After perhaps being behind initially, South Korea's more robust industrial base would probably allow it to surpass North Korea in a nuclear arms competition. As for chemical weapons, Iraq already has demonstrated its superiority over Iran. The relative balance of chemical-weapons capabilities elsewhere is more difficult to gauge based on publicly available information, though differences can be expected.

The increased complexity of military balances in Third World regional confrontations will make it harder to calculate the relative advantage and disadvantage of using force. While such uncertainty might sometimes engender caution, in other situations it could lead to a greater readiness to take military risks or to use conventional force. This might be the case in confrontations in which only one side possesses nuclear or chemical weapons. The greater complexity will also add to national insecurities and fears, while increasing the risk of miscalculation in time of crisis or low-level conflict.

Crisis Instabilities and the Likelihood of Conflict

Armed with nuclear warheads, ballistic missiles would permit rivals in the Middle East, South Asia, the Persian Gulf, and Northeast Asia to inflict devastating damage on each other's cities, people, and military assets. Absent effective defenses, a preemptive strike against those missiles or their warheads would provide the only potential means for a country to avoid or greatly reduce the threat of nuclear devastation. Moreover, at least initially, the missile forces of many countries are likely to be deployed above ground, to be slow to make ready for use, and to be vulnerable to preemption. In this situation, intelligence warning (or a false alarm) during a crisis or conflict that a rival had begun to disperse its missiles or warheads could lead to fear that that country was planning to use its nuclear weapons. Pressures to preempt, to get in the first nuclear blow, would jump considerably.

Moreover, once rival forces had begun to increase their alert rates and prepare for possible use, it might be very difficult to stop the process because standing down from stepped-up readiness would make a side significantly more vulnerable to the opponent's use of nuclear weapons. As the conventional conflict intensified, a use-them-or-lose-them psychology could set in. Further, with increased readiness and preparations for use, the risk of unauthorized use of a nuclear-armed missile or an accidental launch would be considerably increased.

Comparable if not greater crisis instabilities would be likely to characterize confrontations between countries with ballistic missiles armed with chemical weapons. Again, the potential payoffs of preemption could be high, as would be the pressures to use missiles before an opponent had completed preparations to launch its missiles. For a country such as Israel, faced with a Syrian chemical missile threat to its key bases, command centers, and mobilization facilities, or for a country such as South Korea, faced with a comparable threat from North Korea, preemption could appear the best way to blunt such an attack. Countervailing restraints, moreover, would be considerably less because the expected civilian damage of chemical-weapons retaliation would be far less than with nuclear weapons.

Over time, the payoffs of either nuclear or chemical preemption are likely to drop as rivals modernize their forces to reduce their vulnerabilities. This period of transition, however, could be an extended one and would be a time of maximum danger for nuclear or chemical conflict in these regions.

Calculated Use of Chemical and Nuclear Weapons

Chemical or nuclear weapons might be used, moreover, not out of fear that an opponent was about to use them, but in a calculated first strike. The risk of such premeditated use would be highest in situations in which the user alone possessed nuclear or chemical weapons and in which their use was thought to be a last resort to avoid conventional military defeat. But other credible scenarios are also conceivable. Iraq's early use of chemical weapons in its war with Iran, for example, defeated Iranian human-wave conventional assaults. Later Iraqi use, by contrast, served as a force multiplier to make possible Iraqi breakthroughs on the Faw Peninsula in 1987 and helped to shatter Iranian troop and public morale. Absent a threat of effective retaliation, as in the 1991 war, neither internal moral or political restraints nor external obligations dissuaded Iraq from using the weapons.

Though more extreme, there are plausible situations in which a Third World country might use nuclear weapons either as a last resort or for comparable military advantage, especially if it alone possessed them. Israel, for example, probably would be prepared to use nuclear weapons against either Syrian or Egyptian tank concentrations if that were necessary to block a breakthrough and prevent a military defeat. Pakistan could do so as well, though it would have to fear Indian nuclear retaliation. Either Iraq or Israel might consider using nuclear weapons should it become engaged in a war of attrition with Iran or with Syria, respectively. Even the possibility of attacks against cities cannot be precluded— North Korea could use nuclear weapons against vital resupply ports in the South in the event of a new Korean conflict, while Israel might be prepared to attack a city in Syria in retaliation for Syrian chemical-weapons attacks on Tel Aviv.

Heightened Intensity of Conflict

The proliferation of weapons of mass destruction and the ballistic missiles to deliver them will increase the intensity of conflict in Third World regions. Use of nuclear weapons could easily result in hundreds of thousands of casualties, as compared with thousands or tens of thousands of casualties in all but a few prior Third World conflicts. (Though the casualty levels of the Korean War and the Gulf wars were higher, they still were less than the number of potential casualties from the use of nuclear weapons.)

Though less destructive, use of chemical weapons could also lead rapidly to thousands of casualties, especially if civilian populations were targets. The political impact of even limited fatalities from chemical weapons would vary from country to country. For some countries, Israel, for example, even a thousand civilian casualties from use of chemical weapons would be considered a major social and psychological disaster.

New Scope of Regional Conflict

The proliferation of advanced, longer range, high-performance aircraft, ballistic missiles, and other power-projection capabilities already has broadened the potential scope of future regional conflicts. New countries are being brought into old conflicts as active participants. For example, the acquisition of Soviet SU-24 aircraft gives Libyan forces the range to intervene in a future Arab–Israeli war; the Saudi acquisition of CSS-2 missiles and high-performance aircraft without basing restrictions is having a comparable impact.

Ballistic-missile proliferation, with or without nuclear warheads, is changing the scope of conflict in another way. In the future the capital cities and other high-value assets—from Israel's Dimona nuclear-weapons complex to Pakistan's Kahuta facility—will be increasingly vulnerable to attack. Previously aircraft either lacked the range to make such strikes or had only a limited prospect of penetrating air defenses. Moreover, as demonstrated by the 1987–1988 Iran–Iraq war of the cities, in which several hundred ballistic missiles were fired at Teheran or Baghdad, regional rivals are quite likely to take this step.

The proliferation of advanced weaponry is increasing the scope of conflict by tying together previously separate regional rivalries or conflicts. Saudi acquisition of CSS-2 missiles operated by Chinese technicians was seen as a threat by India. One reason for Israel's development of a longer range ballistic missile is to be able to deter a nuclear-armed Pakistan from becoming involved in a future Arab–Israeli conflict. And Indian and Pakistani deployment of nuclear weapons will influence the patterns of future Gulf conflict.

Global Spillovers of Proliferation

The proliferation of ballistic missiles, chemical weapons, and nuclear weapons will pose new threats to U.S. and Soviet allies and friends in the Middle East, the Persian Gulf, South Asia, and the Korean Peninsula, with new calls for superpower involvement in regional conflicts. It will also heighten the difficulties and dangers of power projection, more perhaps for the United States than for the Soviet Union. Access to overseas bases may be affected. Under some conditions the national territories of the Soviet Union and then the United States might even be put at risk by new nuclear-weapon states with missiles.

Regional Threats to U.S. and Soviet Interests and Allies

Many U.S. allies and friends are likely to be threatened. Some particularly dangerous examples include Libyan deployment of ballistic missiles, possibly with chemical warheads, capable of striking targets in Egypt and Israel; Iraqi deployment of nuclear weapons, posing a threat to Israel as well as to Egypt,

Saudi Arabia, and other moderate Arab governments; and North Korean production of nuclear weapons, affecting not only South Korea but possibly Japan.

This intensified threat is likely to lead to requests from U.S. friends and allies for stepped up security ties and in some instances nuclear guarantees. Faced with a hostile Iraqi nuclear force, Saudi Arabia and other regional states would be likely to seek an extension of the U.S. nuclear umbrella—so would South Korea should North Korea's nuclear program not be stopped short of nuclear-weapons production. Alternatively, the threat posed by chemical weapons and ballistic missiles could result in requests for assistance with both passive and active defenses. For its part the Soviet Union may also come under greater pressure to provide security guarantees (e.g., to a Syria faced with an Iraqi nuclear threat or an acknowledged Israeli nuclear capability).

New Costs and Risks of Power Projection

Missile, chemical, and nuclear proliferation in the 1990s will add to the difficulties as well as the potential risks for the United States of projecting power in the Third World to protect traditional allies, friends, and interests. Turning first to missile proliferation, the mobility of naval forces backed up by their organic defensive capabilities (from preemptive strikes against missile bases on warning of attack to fleet missile-defense capabilities) makes it unlikely under many conditions that ballistic-missile attacks with conventional warheads on those forces would inflict significant, if any, damage. But lucky shots are not precluded, and sophisticated explosives and fragmentation warheads could increase the radius of potential damage.

The availability of chemical warheads for ballistic missiles would further enhance the risks and potential costs of projecting military power. Here, too, active and defensive measures, including wearing protective gear and buttoning up naval vessels, could be taken to neutralize the threat of chemical attack. Mobility would make it hard, but not as difficult as with conventional warheads, to target U.S. forces. However, even the possibility of a chemical attack would force changes in U.S. procedures, increase the difficulties facing U.S. forces, and reduce their effectiveness. In regions such as the Middle East and the Persian Gulf, moreover, this reduced effectiveness could be very considerable. Against unprotected troops on the beach or a ship that had failed to take protective measures on warning of an incoming missile, the use of chemical weapons by a Third World country could be devastating. Further, use of chemical weapons would impede cleanup after conventional munitions attacks on bases and ports, while adding to the public havoc and political outcry.

Nuclear proliferation would greatly increase the dangers of projecting power in the Third World. The costs of successful intervention could jump upward to levels previously possible only in clashes with Soviet forces. Although dispersal of naval forces would again provide a partial counter, the greater destructiveness of nuclear weapons means that even a near miss could do very significant damage to sensitive electronic and fighting equipment, if not sink one or more ships. Forces ashore would be even more vulnerable, as would offshore naval forces at anchor or operating in a confined space. Key ports and airfields needed to resupply U.S. or friendly forces (e.g., Diego Garcia in the Indian Ocean or Pusan in South Korea) also would be vulnerable to attack.

Faced with such a small-power nuclear threat, preemptive strikes against missile sites and tactical missile defenses would be one line of active defense. Political constraints and technical difficulties, however, could rule out successful preemption. As for antitactical missile defenses, the effectiveness of ship-borne systems remains untested. The main counter to small-power nuclear threats, therefore, is likely to be the threat of devastating retaliation to the use of nuclear weapons against U.S. forces. The prospects for successful deterrence would be dependent on the elements of the particular situation, including the stakes in the eyes of the new nuclear power; its estimate of U.S. presidential and public willingness to carry out the threat of retaliation; the rationality (or fanaticism) of its leadership; and the risk of accidental or unauthorized small-power nuclear use. In some situations a breakdown of deterrence should not be discounted.

The impact of proliferation on Soviet power-projection capabilities may be different than the impact on U.S. capabilities. Soviet military doctrine and postwar posture have placed less emphasis on naval forces as a key instrument of power projection. Indeed, Soviet naval forces have only very rarely been used militarily since 1945. Greater emphasis appears to be placed on the use of client countries with prestocked equipment (e.g., Libya or Vietnam), which could then serve as bases of operations in a region. For the Soviet Union the greatest vulnerability, therefore, would be airfields and equipment depots in regional countries, rather than its naval forces.

For still other outsiders the proliferation of advanced weaponry will create new direct and indirect risks of several sorts. Merchant ships transiting key war zones, the Straits of Hormuz in a renewed Gulf War, for example, could become targets for land-based missiles. Distant bases for U.S. forces in allied countries might be attacked by extended range or longer range ballistic missiles. The simple threat of such an attack could generate public pressures to revoke such base rights or increase their costs.

Outside Involvement and Regional Conflict

In the past there have been two different patterns of outside reaction to regional conflicts in the Third World. In the wars both between the Arab countries and Israel and between India and Pakistan, outsiders have sought to use their influence quickly to bring the conflict to a close. Political leverage and even threats of military intervention have been used. By contrast, during the 1980s Gulf War between Iran and Iraq, outsiders mostly stood aside. This was so even after the extensive Iraqi use of chemical weapons.

Traditional interests, political ties, and humanitarian concerns are likely to provide incentives for both the United States and the Soviet Union to come to the support of allies or friends facing new threats. At the same time, the potentially greater costs of military involvement in all probability will argue for greater U.S. (if not also Soviet) caution in projecting power in the Third World. This argument for caution, moreover, is likely to be reinforced by the dangers of a U.S.–Soviet confrontation arising out of a proliferation incident. In the Middle East, South Asia, and Northeast Asia the United States and the Soviet Union are tied to different sides in regional conflicts. Action by one superpower to support its ally or friend could lead to a confrontation between the United States and the Soviet Union.

The particular balance between these considerations will depend on the specific situation as well as on the particular leaders in Washington and Moscow. More likely than not, support for traditional friends and allies in Third World regions will sometimes be tempered by concerns about the increased risk of confrontation as well as the dangers of military involvement. In at least some situations the choice may be to stand aside militarily. If there is a risk of nuclear-weapon use, or in situations in which both regional opponents, for example, Iran and Iraq, are considered international outcasts, the likelihood is greater that the two superpowers will stand aside.

New Threats to National Homelands?

Until now, the proliferation of advanced military capabilities in Third World regions has not posed a military threat to the national territories of the United States and the Soviet Union. With several important exceptions, this is likely to remain the rule in the coming decade.

By the end of the decade, technological advances in the missile programs of some countries (e.g., Israel, Brazil, or India) could result in their deployment of longer range ballistic missiles. But with the exception of Israel, it is difficult to conclude that they would have political and military incentives to target either superpower's homeland. On the contrary, nearly all of the countries that are acquiring chemical- or nuclear-weapons or missile capabilities are doing so because of long-standing disputes with their regional rivals. As for Israel, public evidence suggests that it is seeking to develop longer range ballistic missiles capable of carrying out a nuclear strike on Moscow. (In this regard the greater distance of the United States from the most plausible proliferation hot spots makes the United States relatively less vulnerable than the Soviet Union to ballistic-missile attacks.)

Of greater concern is the threat of terrorist access to chemical or nuclear weapons, or chemical- and nuclear-weapons materials, in Third World countries. The theft of a weapon or materials could occur either with or without insider help. In an era of state-sponsored terrorism, moreover, provision of chemical weapons—and perhaps even a nuclear weapon—to a terrorist group cannot be discounted completely. With additional chemical- and nuclear-weapon countries, this risk of loss of control will increase. It will also increase as the next nuclear-weapon states begin to add to their stockpiles of materials and weapons. In both cases numbers alone will provide more opportunities to acquire materials, if not weapons, by theft or corruption.

A PROLIFERATION CONTAINMENT STRATEGY

A comprehensive approach to proliferation policy is required to meet the challenges of the 1990s. Continued measures to check the further spread of nuclear and chemical weapons as well as ballistic missiles comprise one aspect of that approach. Greater caution also is in order with regard to potentially destabilizing sales of advanced conventional weaponry. But it will be necessary to complement strengthened efforts to slow further proliferation with new measures to contain its regional consequences and global spillovers. In effect, future U.S. policy should think in terms of containing the scope, pace, and consequences of an overall process of proliferation.

A full discussion of such a comprehensive proliferation containment policy exceeds the scope of this essay. What follows, instead, is a kaleidoscope of possible policies to illustrate the types of measures and the breadth of approach required in the respective areas of nuclear- or chemical-weapons or missile proliferation. We consider first measures to check further proliferation and then measures to contain the regional consequences and global spillover effects of proliferation.

Checking Further Proliferation

Preventing the further spread of nuclear weapons has been a long-standing U.S. goal. With other countries' support, measures have been taken to control nuclear-related exports and to make it technically harder for a country to acquire nuclear weapons; to lessen political and security incentives to build a nuclear-weapons capability; and to put in place global nonproliferation institutions and to strengthen a norm of nonproliferation. In 1987 the United States, the United Kingdom, Japan, France, the Federal Republic of Germany, Italy, and Canada reached agreement on a parallel effort to check the spread of ballistic and cruise missiles and missile technology—the Missile Technology Control Regime

(MTCR). Stepped up efforts to contain the spread of chemical weapons have emphasized export controls (under the so-called Australia Group of Western exporters) and the negotiation of a complete and total chemical-weapons ban. New initiatives of several sorts can buttress these policy thrusts.

Ensuring Effective Nuclear Export Controls

Nuclear export controls have helped to slow the pace of nuclear proliferation and to increase the difficulties of acquiring nuclear weapons. Controls have been especially effective over items (e.g., nuclear reactors, complete reprocessing plants, and nuclear materials) directly usable for nuclear activities. Looking ahead, stepped up efforts are needed to put in place multilateral controls over so-called dual-use exports, items with both legitimate nuclear and nonnuclear uses.

In addition, the creation of a single internal European market in 1992 may make it easier for countries seeking nuclear weapons to circumvent export controls. It could do so either by weakening the legal status of regulations in individual European countries (in an era of open borders) or by facilitating transshipment from countries with well-policed export control regulations to countries with poorly policed ones. Discussions with European suppliers and possible steps by the European Community to give export controls such as the Nuclear Suppliers' Guidelines the status of Community law would be one response.

Economic and political change throughout Eastern Europe may also present new problems for nuclear export controls. Soviet influence can no longer be relied on to encourage caution in Eastern European governments. For some East European countries, nuclear and nuclear-related exports could be an attractive source of hard currency earnings, including possible sales to countries of questionable nonproliferation status. Before new regulations and practices are established, moreover, the types of entrepreneurial middlemen that proved so important in Pakistan's access to sensitive and dual-use exports for its nuclear-weapon program could also spring up in these countries. Bilateral discussions with East European countries on nonproliferation matters, therefore, should be added to the U.S. diplomatic agenda.

Defusing Regional Nuclear Proliferation Hot Spots

New diplomatic initiatives are also required to bring the combined political and economic influence of many countries to bear in attempting to contain nuclear-weapons activities in South Asia, the Middle East, and Northeast Asia. Both India and Pakistan appear reluctant to cross the nuclear-weapon threshold and begin an open nuclear arms race in the region. The United States, the Soviet Union, Japan (a key economic partner of India and Pakistan), and still other countries should use their influence to reinforce that reluctance. Additional Indo-Pakistani confidence-building measures (e.g., visits to sensitive facilities, pledges not to detonate a nuclear explosive, and pledges not to build and deploy nuclear weapons) might be encouraged to complement their mutual commitment not to attack each other's nuclear facilities.

In the Middle East the United States needs to put the nuclear issue on its agenda with Israel to make clear the importance of nuclear restraint. Comparable Soviet diplomatic initiatives to contain any growing Syrian nuclear ambitions are in order.

Still elsewhere, Soviet diplomacy has been brought to bear (so far unsuccessfully) to restrain North Korea and convince its leadership to follow through on its commitment to accept international inspection of its nuclear facilities. This needs to be tried again. Chinese support should be solicited as well. For its part the United States should avoid radical reductions of the U.S. combat military presence in South Korea lest that provoke South Korea to take out of mothballs its earlier nuclear-weapon program.

Extending the NPT

One of the main nuclear nonproliferation challenges ahead will be extension of the NPT, either indefinitely or for a fixed period or periods, when it comes up for renewal in 1995. Prospects now are good for extension for an extended period, perhaps another twenty-five years. There is general agreement that the NPT has added to the security of its parties and helped to foster the peaceful uses of nuclear energy. Expected progress in reducing nuclear and conventional arms will demonstrate progress in meeting the NPT's disarmament goal and probably will outweigh most countries' disappointment over the lack of a ban on nuclear testing.

But extension is not ensured. Possible shocks could make it harder to renew the treaty or could even derail the process. Prominent dangers include an arms-control breakdown; a withdrawal from the NPT or clear-cut violation of its terms by North Korea, South Korea, Iraq, or Iran; or open nuclear-weapon deployments by India, Pakistan, or Israel. A decision by South Africa to adhere to the treaty, but without good-faith steps to place all of its nuclear activities under international inspection, would also be very damaging to the NPT's credibility. Efforts to amend the NPT to add new obligations—well intentioned or not—could backfire as well.

Bringing the Soviets into the Missile Control Effort

Approached by Western countries soon after the April 1987 announcement of the MTCR, the Soviet Union has declined to adhere to this regime. Instead,

Soviet officials have called for a new multilateral regime with still tighter controls. As proposed, these new controls would lower the MTCR's thresholds of control from missiles with a range of 300 kilometers and a payload of 500 kilograms to missiles with a range of 100 kilometers and a payload of 200 kilograms.

In 1990, however, the Soviet Union indicated it would abide by these controls on exports of missiles and missile technology; this is a most welcome next step. Not only has the Soviet Union been the primary exporter of missiles in the past, but future Soviet missile exports (e.g., advanced *Scuds*, the SS-21, and still others) could increase the range, reliability, and overall threat posed by Third World missiles.

Chinese acceptance of missile export restraint is also important, but not to the same extent as that of the Soviet Union because of lesser Chinese export capabilities. It appears highly unlikely that the current Beijing regime will formally agree to control missile exports. More likely, periods of restraint will alternate with pursuit of missile exports. The United States and other governments will need to keep a close watch on Chinese activities to use their influence to head off widespread missile sales in unstable regions.

Still other missile suppliers are likely to emerge in the 1990s (e.g., North Korea, Brazil, Argentina, Israel, India, and Pakistan); some of these already have exported or shared technology (e.g., Israel, Egypt, Argentina, and North Korea). Most of these countries, however, would be little influenced by Western—or for that matter, Eastern—calls for export restraint. Their capability to supply modern, reliable longer range systems, however, probably will be limited. Some of them, such as India, may not choose to export at all. Nonetheless the existence of these new suppliers makes clear again that missile nonproliferation can buy time and affect the characteristics of the future Third World missile threat, but it cannot block it altogether.

Scrutinizing Exports of Advanced Conventional Arms

Across-the-board cutbacks of conventional arms exports are neither desirable nor feasible. Under certain conditions, conventional-arms exports can serve important U.S. national security, political, and economic interests. Besides, without parallel action by the Soviet Union and Western European arms exporters, unilateral U.S. restraint will have little impact. But many years' experience indicates that these countries can be expected to reject calls for major new arms sales restrictions.

Within these limits, however, U.S. exports of advanced conventional arms need close scrutiny on a case-by-case basis to prevent arms sales that would be likely to result in greater regional instability. Sales of aerial refueling and airborne warning and control systems that would provide Pakistan with an enhanced nuclear strike capability against India would be one example. Discussions with other arms suppliers could also focus on specific exports, such as Soviet exports of advanced aircraft to North Korea or to Nicaragua, that would heighten regional tensions and add to the risk of conflict.

Wrapping Up a Complete and Total Chemical Weapons Ban

On the chemical-weapons nonproliferation front, export controls can buy time. But a top priority needs to be given to the successful negotiation of a complete and total chemical-weapons ban. Only such a global ban holds out any hope of bringing the problem of chemical-weapons proliferation under control. Such a ban would help to reverse self-fulfilling fears of runaway chemical-weapons proliferation; would strengthen export control efforts by providing them with an international legal foundation; would preclude or, in the event of violation, significantly increase the difficulties of producing and deploying chemical weapons in treaty adherents; and would provide a political and moral foundation for the use of sanctions up to military force to block chemical-weapons programs.

A complete and total chemical-weapons ban has real risks. Some verification uncertainties will be unavoidable, particularly in detecting the undeclared clandestine production of chemical-weapons agents or weapons. Under some conditions, the use of illegal chemical weapons by the Soviet Union or a Third World country against U.S. forces could be militarily significant. Some key parties, especially in the Middle East, may prove reluctant or unwilling to join.

However, with the declining Soviet conventional military threat in Europe and the breakup of the Soviet Eastern European empire, the military (but not political) significance of undetected Soviet chemical weapons has greatly dropped. Though potentially costly, the use of illegal chemical weapons by a Third World country against U.S. forces would be unlikely to preclude successful military action. U.S. counters other than chemical-weapons retaliation exist.

At the same time, it is important to begin thinking about how to use diplomatic, political, military, and economic influence to support the widest possible adherence to a chemical-weapons ban. Particular attention could be paid to assistance that might be provided to parties to the treaty were they to find themselves threatened with chemical weapons by a nonparty or by a treaty violator. The prospect of assistance could be a key selling point. In turn, more thinking is in order about possible sanctions against violations, including the use of chemical weapons. This, too, would help to reassure countries that would want to join, while lessening the payoffs of staying out of the treaty—let alone using chemical weapons—for some of the most troublesome potential holdouts.

Containing Regional Consequences and Global Spillovers

Future U.S. strategy for dealing with nuclear- and chemical-weapons and missile proliferation increasingly will need to include measures to contain its regional consequences and global spillovers. Several different sets of containment measures may be required, varying in part with the specific proliferation threat. Broadly defined, these include policies and initiatives to establish proliferation firebreaks, to lessen the threat to U.S. allies and friends, to strengthen command and control as well as physical security over new nuclear arsenals, to adapt U.S. power-projection capabilities to a changed military environment, and to control the risk of a U.S.–Soviet confrontation arising out of a proliferation incident.

Establishing Proliferation Firebreaks

Particularly in the case of nuclear proliferation, efforts will be required to establish proliferation firebreaks even after some initial proliferation has occurred. For example, in South Asia, political influence could be used to convince India and Pakistan not to advance from future unacknowledged to open nuclear-weapon programs. The types of confidence-building measures discussed earlier could play a role in that effort. Both sides' agreement to a regional nuclear test ban would help to contain the magnitude of the threat posed by each side's nuclear-weapons capability to the other. Taking these measures a step further, arms-control negotiations to limit each side to a minimum deterrent nuclear force might be pursued.

Closely related, under some conditions, limits could be encouraged on missile deployments by Third World countries. Israel has yet to deploy longer range missiles capable of striking Moscow, a major and potentially very destabilizing threshold. In South America, arms-control negotiations between Argentina and Brazil could seek a de facto agreement to export but not deploy more than a token force of newly developed missiles. Verification measures comparable to those now accepted by the United States and the Soviet Union could be added to reassure both countries that nuclear warheads had not been fielded on such missiles.

Security Guarantees for Threatened Allies and Friends

New or more explicit U.S. security guarantees to close allies or friends threatened by new nuclear powers are another aspect of an overall containment strategy. Even with the greater dangers of such guarantees, the economic, strategic, or political importance of some U.S. allies will call for such a commitment. Extending the U.S. nuclear umbrella to a Saudi Arabia faced with a newly nuclear Iraq, for example, is one possible example. An explicit commitment to Seoul of a U.S. retaliatory response in the event of nuclear use by North Korea could be another. Assuming a decision to provide such a nuclear security guarantee, important practical details would need to be worked out concerning what to say about it, what type of retaliation to threaten (nuclear or devastating advanced conventional), and what other steps to take to make it credible.

Similarly, in the event of more extensive chemical-weapons proliferation, U.S. security interests might justify a decision to offer a U.S. guarantee to retaliate on behalf of a country (e.g., Saudi Arabia). The prospect of such retaliation may be particularly important to prevent future unraveling of a chemical-weapons ban in the event that there are holdouts or violators. Assuming U.S. adherence to such a ban, this would require, of course, alternative means of retaliation than chemical weapons, quite possibly the use of advanced conventional weaponry.

Defensive weapons assistance also could help counter future chemical-weapon threats to U.S. allies and friends. This might even include for some situations the rapid deployment of tactical ballistic-missile defenses, such as the *Patriot* batteries deployed in the 1991 conflict. An ability to provide active defenses against regional ballistic-missile threats might be needed, as well, to convince countries to continue to provide bases and access for U.S. military forces.

Strengthening Control and Security

The costs and benefits will need to be carefully weighed of helping Third World countries to tighten control and security over new missile forces as well as over nuclear weapons and materials. Such support could range from the provision of advice and paper studies up to the transfer of control and security technology.

Even a consideration of such assistance is likely to be controversial. Assistance would lessen an important disincentive to acquiring nuclear weapons (fear of loss of control). It also could run counter to the NPT obligation not to assist other countries to acquire nuclear weapons. However, steps to strengthen control and security would lessen the danger that terrorist groups would gain access to nuclear materials or weapons. Assistance would also reduce the risk of the unauthorized use of missiles or nuclear weapons in a crisis or conflict, as well as the risk of a nuclear or a missile accident. This could be especially important given the probable crisis instabilities in the most likely future regional confrontations.

Striking the right balance between these conflicting considerations must probably rely on a case-by-case determination. The timing of assistance is likely to be important: not so soon as to stimulate proliferation, but not too late to matter.

Adapting U.S. Defense Posture for Third World Missions

Some modifications at the margin will certainly be needed in U.S. defense planning and the U.S. posture for power projection in Third World regions. Doctrine, training, personnel, tactics, equipment, intelligence, command and control, and rules of engagement could be affected by missile, chemical, and nuclear proliferation. Consider just a few examples. The active defense of U.S. Navy ships as well as fixed assets such as naval bases and amphibious forces against a Third World ballistic-missile threat will depend on the antitactical ballistic-missile capability of the *Aegis* cruiser. But ensuring its effectiveness against that threat could warrant its prior testing in that role. Or rules of engagement might have to be developed that would permit rapid preemptive response by the local commander on warning of the possible use of ballistic missiles (with or without nuclear or chemical warheads) against U.S. forces, allies, or friends. Intelligence collection concerning a Third World opponent's nuclear- and chemical-weapons and missile capabilities, as well as command and control to make use of that intelligence, would have to keep pace.

Reducing the Risk of U.S.–Soviet Confrontation

Discussions between the United States and the Soviet Union about the risk of unintended military confrontation arising out of a proliferation incident are another part of a more comprehensive proliferation containment strategy. These discussions could grow out of current informal military-to-military contacts, first initiated by former Chairman of the Joint Chiefs of Staff Admiral William Crowe. Or a more political venue could be sought, building on current U.S.–Soviet bilateral discussions.

As a start, any such talks might exchange views on each side's perception of threats to its security stemming from missile, chemical, or nuclear proliferation problems. Possible scenarios for proliferation-related incidents that could result in a U.S.–Soviet confrontation might be explored. Measures could then be examined to lessen that risk, including provisions for rapid communication between theater commanders, rules of the road for specific regions, and common actions to restrain allies and friends.

Exploring Missile Defenses

At least initially, the threat to the U.S. homeland posed by future missile and nuclear proliferation is likely to be very limited. This would be less so for the Soviet Union, particularly if Israel moves to deploy longer range ballistic missiles. Both Moscow and Washington, therefore, may find it useful to explore responses to this threat within the broader framework of continuing discussions of strategic missile defenses. For the most part, the purpose would be to determine how each side viewed the threat and the potential role of limited missile defenses in containing it. If needed by a future Third World threat, later negotiations concerning new deployments of a limited area defense—now precluded by the 1972 Antiballistic Missile Treaty—could build on those talks.

CONTAINING THE PROLIFERATION THREAT

During the 1990s the military characteristics of regional confrontations in the Middle East, the Persian Gulf, South Asia, Northeast Asia, and even South America could change considerably. Further proliferation of advanced weaponry, especially ballistic missiles, chemical weapons, and nuclear weapons, is all but unavoidable. This will have far-reaching implications for regional conflict and global order. In light of those consequences the United States needs to keep up both its traditional nuclear and its recently stepped-up chemical and missile nonproliferation efforts. These can slow that process of proliferation and possibly check its most extreme manifestations. However, a new and more comprehensive proliferation strategy is also demanded. This strategy must blend new measures to contain proliferation's regional consequences and global spillovers with traditional policies. It is none too soon to get on with this job.

NOTES AND REFERENCES

1. Remarks before the Council on Foreign Relations, December 12, 1988.
2. Elisa D. Harris, "Stemming the Spread of Chemical Weapons," *The Brookings Review* 8 (Winter 1989–1990), pp. 39–41.
3. *The Washington Post*, March 31, 1989.
4. Harris, "Stemming the Spread."
5. *The Washington Post*, July 29, 1989.
6. *The New York Times*, March 23, 1988.

11

Military and Civilian Uses of Space
Lingering and New Debates

Michael Krepon

INTRODUCTION

President Ronald Reagan's Strategic Defense Initiative (SDI) and his administration's attempts to resume tests of antisatellite weapons (ASATs) dominated our national security debates over America's role in space during the 1980s. In contrast, the most dramatic space-related event during the decade—the loss of the space shuttle *Challenger* and its crew—did not generate a wide-ranging, divisive debate over America's future in space. Instead, this tragedy focused public and expert attention on why the accident occurred, how to operate the shuttle safely, and ways to diversify America's access to space in the future.

The SDI and ASAT issues will not fade away during the 1990s, but there are reasons to expect them to be less divisive in the years to come. Increasingly over the next decade, public attention is likely to turn to a new range of less contentious issues generated by the diffusion of satellite technology and its multiple impacts on multilateral and public diplomacy, media operations, and crisis decision making.

Michael Krepon • The Henry L. Stimson Center, Washington, D.C. 20036.

LINGERING DEBATES FROM THE 1980S

President Reagan's surprise unveiling of SDI prompted the harshest American national security debate since the second Strategic Arms Limitation Talks (SALT II) ratification struggle in the twilight of the Carter administration. The 1980s debate was about a fundamental choice between maintaining the Anti-Ballistic Missile (ABM) Treaty or casting it aside in the pursuit of effective defenses against ballistic-missile attack.

Much of this same ground was covered during the latter half of the 1960s. Then, as during the 1980s, alarms were raised about impending Soviet efforts to deploy nationwide defenses that would place the United States at a great strategic disadvantage. Then, too, some strategic analysts and leading figures in think tanks, industry, and the executive and legislative branches said defenses could work effectively at affordable costs. But in 1972 the effort collapsed with the signing of the ABM treaty. No less a supporter of strategic defenses than Edward Teller accepted the Nixon administration's decision:

> ... the ABM Treaty which is up for ratification ... merely recognizes political realities. We must make a choice if we indeed care for the American people. We might decide to spend a minimum of an additional $100 billion for defense in the next five years, and thus build a full-scale antimissile defense system. No great gift of prophecy is needed to know this will not happen. The only alternative is to ratify the treaty.[1]

It was not just fiscal and political realities that militated against nationwide ballistic-missile defenses during the Nixon administration: ballistic-missile-defense technology was then also far too primitive to prevent radar blackout or to permit discrimination between warheads and decoys during attacks. False claims to the contrary made during the heat of debate from 1969 to 1971 have subsequently been thoroughly discredited.

New technological developments, recurring fears about Soviet strategic objectives, and President Reagan's personal vision of a peace shield that would make nuclear weapons impotent and obsolete combined to generate a new push for strategic defenses in the 1980s. But many of the old arguments continued to bedevil the advocates of SDI. The Reagan and Bush administrations' criteria for deployment—survivability and cost-effectiveness at the margin—remain formidable barriers. Despite considerable progress in many of the technologies needed for effective strategic defenses, Bernard Brodie's observation, "[T]here is not much solace in raising the enemy's requirements if he is still able to meet them,"[2] continues to be applicable.

The high technical and political hurdles facing SDI deployments will be raised further by financial constraints facing the Pentagon throughout the 1990s. Without tenacious support from within the Department of Defense, SDI faces a most uncertain future. The opportunity costs of deploying robust strategic defenses will inevitably erode bureaucratic support for SDI within the Pentagon because traditional military and bureaucratic preferences point to offensive strategic modernization programs rather than defensive deployments.

The large research, development, and test budgets allocated to SDI will also increasingly jeopardize necessary research and development in other areas. By 1989 the SDI program had already garnered more research funds than those allocated to the U.S. Army; as SDI funding requirements increasingly affect important defense research projects in other areas, opposition to the program will inevitably grow, within as well as outside the Pentagon.

Another important development that bodes ill for the deployment of thoroughly effective strategic defenses during the 1990s is continued improvement in U.S.–Soviet relations. President Richard Nixon was unable to persuade the Senate to deploy nationwide strategic defenses at a time when the Soviet Union was concurrently engaged in unprecedented offensive and defensive deployments. President George Bush is unlikely to fare any better when, under the leadership of Mikhail Gorbachev, the Kremlin has clearly embarked on a course leading to better relations with the United States.

Given the repeated nature of the surprises emanating from Moscow and the fluidity of the political situation there, no one can predict the future of superpower relations with supreme confidence. Nevertheless, one critical rationale for improved relations—the need to rebuild the Soviet economy—exists independently of the current Kremlin leadership. A return to a foreign policy based on threat and military buildups will forfeit the goodwill that Gorbachev has accumulated, without helping the Kremlin to strengthen its economic performance.

These conditions point to further improvements in U.S.–Soviet relations unless political turbulence within the Soviet Union or in Eastern Europe short-circuit the process. Improved bilateral relations make it extremely difficult to argue for strategic defenses against the Soviet threat. Instead, public expectation and political convention dictate arms reduction agreements to codify the improved superpower relationship.

In the first instance the focus of public attention will be on a Strategic Arms Reduction Talks (START) agreement that could cut the Kremlin's most potent intercontinental ballistic missiles (ICBMs) by 50 percent, with additional cuts in Soviet multiple independently targeted reentry vehicle (MIRV), land-based missiles. Reductions of this magnitude roughly coincide with the Joint Chiefs of Staff's stated objectives for the effectiveness of first-phase deployments of a strategic defense system in drawing down an attacking force. Advocates of SDI are faced with a cruel irony: Ballistic-missile defenses can potentially become far more effective with negotiated reductions in strategic arsenals, but the successful implementation of such reductions undermines the case for thoroughly effective defenses.

The ratification of a START agreement may also be linked politically to strategic modernization programs that generate domestic confidence in the survivability of U.S. strategic forces remaining after reductions occur. The prospect of a START agreement may help, therefore, to resolve more than a decade of divisive debates over whether and how to remove some U.S. intercontinental ballistic missiles from their silos. More likely, the perceived threat of a Soviet preemptive attack may be so diminished, and budgetary constraints so intractable, that all mobile basing options may be postponed for the near future.

In either case, prospects for thoroughly effective SDI deployments do not appear bright. A solution to the long-standing problem of ICBM survivability will further erode the case for SDI. Funding for mobile ICBM basing arrangements will inevitably compete with other strategic weapon programs, including strategic defenses. Moreover investments in survivable nuclear forces undermine the arguments that SDI deployments are essential to complicate attack plans, thereby reinforcing deterrence, or to provide protection against surprise attack.

These considerations, together with President Bush's more level-headed view of SDI's prospects and continuing support by congressional majorities for the ABM treaty, point to less combustible debates over strategic defenses than during the Reagan years. Still, President Bush will not be able to avoid controversy over SDI because he must either decide to commit the United States to some form of deployment and to structure test programs accordingly or to temporize and to withhold a deployment decision for the remainder of his administration. Any decision will alienate at least one, highly vocal political constituency.

If President Bush opts for a stretched-out research, development, and test program instead of near-term SDI deployments, he will face withering fire from the right wing of his party, who are most concerned about American vulnerability to Soviet attack and the Kremlin's ability to break out from ABM treaty constraints. Such a decision would also leave Bush's successor poorly positioned to opt for deployments if, as SDI supporters fear, funding for a long-term SDI research program declines over time.

Alternatively, President Bush could energize other political constituencies by opting for SDI deployments during the 1990s. A deployment decision can be keyed to ground- or space-based interceptors or some combination of the two. Space-based interceptors involve new techniques for old-style concepts. Much technical development and testing needs to be performed in this area, making a near-term deployment decision fraught with risk, especially because the United States does not now have in hand the technology to discriminate between warheads and decoys. Space-based defenses are likely to generate countermeasures because they pose the greatest potential threat to each side's ballistic missiles. For the same reason, they make significant arms reduction agreements more difficult to negotiate. Defenses in space are also barred by the ABM treaty, ensuring a bitter—and perhaps losing—battle with the Congress, unless the Soviets themselves resume the testing of such weapons in space.

The idea of limited defenses on the ground to provide some protection against an accidental or unauthorized launch of ballistic missiles—the Accidental Launch Protection System (ALPS) proposal initially suggested by Senator Sam Nunn—represents the middle ground in contentious debates over SDI, but it may not be able to garner sufficient political support. Most SDI enthusiasts will support any deployments as an opening wedge to full-scale defenses—the very reason why many Star Wars skeptics will oppose ALPS. The latter would likely form an awkward coalition with some SDI supporters who will fight the diversion of program funds into older technology with limited utility.

A variety of arguments can be lodged against ALPS, such as the system's inability to detect decoys from incoming warheads, and the high cost of even modest defenses. But if ALPS deployments are strictly limited and take place in the context of improved superpower relations marked by nuclear arms reductions, there would be little incentive for either side to employ decoys or other countermeasures. And the price of such a limited insurance policy would be far below the thirty billion dollars the Soviets claim was the cost of cleaning up after the accident at Chernobyl.

Other contentious debates would flow from an administration decision to deploy land-based defenses during the 1990s. A key question is whether such deployments could comply with the ABM treaty, and if not, whether treaty amendments would work in the net U.S. interest, assuming they could be successfully negotiated. No defensive deployment can hope to protect the United States or its allies from all conceivable threats, and modest defensive deployments are particularly susceptible to arguments that they do not provide sufficient insurance for the advertised cost. The alternative, however, is to have no insurance protection against even the most modest threat, even though the resulting damage can be quite costly and traumatic.

A decision to deploy ground-based interceptors would also fuel debates over the resumption of ASAT tests, because these interceptors can be employed against satellites as well as missiles. During the 1980s, ASAT debates were overshadowed by more contentious battles over SDI, but both resulted in highly unusual and informal tacit agreements. On SDI the legislative branch agreed to provide over $17 billion for the program, while the executive branch grudgingly agreed to adhere to the ABM treaty. The research funding level, while generous by most standards, was still well below that requested by the executive branch. Nevertheless, Reagan administration officials clearly understood from well-placed sources on Capitol Hill that congressional support for SDI funding would drop dramatically in the event that the ABM treaty were abrogated or the more convenient but fallacious broad interpretation of that treaty were to guide Star Wars tests.

The tacit agreement on ASATs that evolved during the Reagan administration was even more unusual because the parties to this informal compact were the United States Congress and the Kremlin. After the Kremlin initiated a unilateral moratorium on ASAT tests in 1982, congressional majorities responded by blocking U.S. antisatellite tests against targets in space until the president certified the resumption of Soviet testing. Congressional initiatives in this area were fanned not only by Soviet restraint, but also by the Reagan administration's clear desire to acquire new ASAT capabilities.

The executive branch's disinterest in formal ASAT arms limitations was succinctly demonstrated in a 1984 report to the Congress that failed to identify a single formal agreement that would not leave the Soviet Union with a significant and destabilizing advantage.[3] In effect, the Reagan administration argued that no formal agreement could cover all potential ASAT threats, and that therefore no such agreement was worth pursuing. Congressional majorities rejected this exacting standard, turning instead to partial measures.

The resulting de facto superpower moratorium on ASAT testing against targets in space is a testament to popular opinion that desires to keep space free of weapons testing and deployment, even when these activities are not prohibited by existing treaties. As long as the Kremlin continues its testing moratorium, efforts by the Bush administration to revive and restructure U.S. ASAT programs are unlikely to alter public or congressional views in any appreciable way. Under these conditions, future ASAT debates are likely to revolve around U.S. testing of ground-based interceptors permitted by the ABM treaty. Those who argue that such tests unwisely open up a competition in ASAT capabilities will have the burden of explaining why Soviet ground-based interceptor tests lack strategic significance while comparable U.S. tests could have dire consequences.

NEW DEBATES FOR THE 1990s

A number of new public debates over space-related issues will take place in the 1990s, especially over how America's limited resources for activities in space can best be applied during an extended period of budget stringency. Under different circumstances the United States might be able to commit enough resources to undertake costly, long-term planetary explorations as well as an extensive study from space of planet Earth and the environmental ills facing it. Instead, American political leaders will need to establish priorities between and within these two broad choices.

Another series of debates will arise out of the spread of observation satellite technology during the 1990s, including the role and impact of the media in utilizing images from space, how information from space can affect U.S. military operations and crisis decision making, and whether America should help or hinder multilateral verification efforts using this technology. None of these questions are as politically explosive as the SDI or ASAT issues. Instead, our new space-related debates will have much in common with many other public policy choices facing American political leaders: all are exceedingly complex and obscure and highly resistant to simple solutions.

OPEN SKIES: THE POLICY ISSUES AND DEBATES

President Bush's resuscitation of the idea of open skies may lead to many positive applications if the Kremlin's long-standing resistance to greater openness by means of aerial surveillance can be surmounted. This essay, however, will be confined to the implications of open skies by means of satellites rather than aircraft.

The idea of open skies predated the first launch of satellites. President Dwight D. Eisenhower's memorable proposal at the 1955 Four Power Summit in Geneva called for the trading of blueprints of military establishments and reciprocal aircraft overflights to guard against the possibility of surprise attack. Nikita Khrushchev's rejection of the open-skies plan as a cover for espionage and the subsequent initiation of U-2 flights over Soviet territory symbolized the unilateral nature of subsequent superpower efforts to collect information about force structure and disposition by means of aircraft and reconnaissance satellites.

Bit by bit, superpower and multilateral efforts at sharing information, cooperating in its collection, and opening national facilities to foreign inspection took their place beside unilateral efforts to gather data by national technical means (NTM). Joint U.S.–Soviet efforts along these lines began modestly with a one-sided exchange of data on strategic offensive forces in the 1972 SALT I interim agreement, as well as cooperative measures to assist NTM in monitoring the terms of the agreement. More intensive cooperative arrangements were negotiated, but not implemented, in the long-unratified Threshold Test Ban and Peaceful Nuclear Explosions treaties; the unratified SALT II treaty's cooperative measures were only partially implemented. The major breakthroughs in cooperative superpower verification, the Intermediate-Range Nuclear Forces (INF) Treaty and the Joint Verification Experiments on nuclear testing, awaited the advent of Mikhail Gorbachev and President Ronald Reagan's second term in office.

In several respects, cooperation in multilateral verification efforts proved easier to achieve, beginning with the nonthreatening inspection provisions associated with the 1959 Antarctic Treaty and the important access granted to International Atomic Energy Agency inspectors under the Treaty on the Nonproliferation of Nuclear Weapons, signed in 1968. Again, a major breakthrough occurred during President Reagan's second term, with the 1986 agreement by all signatories to the Conference on Disarmament in Europe (CDE) to allow inspections of certain military exercises without a right of refusal. Almost thirty years after the open-skies speech, the INF treaty's verification provisions and the modest transparency permitted under the CDE accord provided a belated confirmation of President Eisenhower's effort.

Others have sought a fuller realization of open skies. A French initiative at the United Nations to establish an International Satellite Monitoring Agency

(ISMA) has received considerable attention and support, as have concepts for more limited regional monitoring arrangements. The Canadian government has proposed a *Paxsat* to help monitor future multilateral agreements governing ground forces and space arms control, while the government of Sweden has announced its readiness to become actively involved in a multilateral consortium to launch and operate a verification satellite with resolution capabilities between one and a few meters. (The Canadian proposal for a radar satellite would necessarily have somewhat poorer resolution.) If any of these proposals are realized, many new peacekeeping and monitoring applications will arise.

Whether or not international monitoring efforts from space can be implemented during the 1990s, activities on the ground are becoming more visible as a result of commercial observation satellites. The French government's decision to launch Satellite Pour l'Observation de la Terre (SPOT), with its ten-meter panchromatic images, has had multiple impacts among states with satellite capabilities. The Kremlin has announced its readiness to sell even higher quality images for hard currency, while the Reagan administration declared its intention not to allow U.S. commercial space ventures to be placed at a disadvantage to foreign competition.

Another flurry of activity can be expected in 1992 after the French launch *Hélios*, an intelligence-gathering satellite that builds on SPOT's capabilities. Spain and Italy will be partners in the *Hélios* program, prompting other nations to assess their future needs for pictures from space and how those needs can be met. Other states will no doubt follow the French lead, launching observation satellites during the 1990s, for commercial, civilian, or military purposes. The availability of these images and the inevitable improvement in their resolution and turnaround time will mean that foreign governments, independent experts, and the news media will have more information at their disposal to use for constructive or mischievous purposes. The diffusion of satellite technology also means that a new set of space policy issues will enter the public arena.

THE ROLE OF THE MEDIA

Domestic arguments over commercial observation satellites initially revolved around the potential operation of a *mediasat* and the conflicts that might result between the press, First Amendment rights, and the executive branch's right to impose prior restraint on the release of information that could adversely affect national security.[4] For a variety of reasons, expert debate over a mediasat dedicated to gathering information from space has abated. First, it is unrealistic to expect a single satellite to be properly positioned during times when U.S. government officials are likely to be most sensitive about their operations. Second, the economics of a single satellite dedicated to news gathering—let alone a constellation of them—appear prohibitive, at least in the near future. Third, foreign ownership of commercial observation satellites makes domestic legal constraints on the utilization of pictures from space somewhat beside the point.

As a result of these factors, future debates over pictures taken from space are likely to revolve not around a mediasat, but around the media's periodic use of imagery taken by commercial observation satellites. The key issues in this regard are not the press's First Amendment rights, but the potential impacts of the media's use of images from space on public opinion and on decision making during crises.[5] Televised pictures can have a powerful impact on public opinion during telescoped or extended crises, whether they be taken by camera operators on the ground (as was the case during the Iranian hostage crisis) or from reconnaissance aircraft, as during the Cuban missile crisis.

Images from space could similarly evoke strong national sentiments, especially when pictures from the ground or from aircraft are not available. The misrepresentation of these images, or their ill-timed release by state-controlled or Western media, can place additional pressures on national leaders and shorten the time frame within which decisions must be made. These potentially adverse impacts will be felt mostly by political leaders in the West because their counterparts in more closed societies will be less susceptible to public pressures, especially during crises, when the mass media can be expected to close ranks behind official government positions.

The spread of satellite technology and the greater worldwide availability of high-quality imagery is therefore not an unqualified boon to the West, although the initial burden of reacting to SPOT imagery has fallen primarily on the Kremlin. This was to be expected, given the relatively closed nature of Soviet society and the media interest in stories about previously inaccessible places depicted by these images.

More surprising was the way that Gorbachev and his allies used commercial satellite images to further their policy of *glasnost* and to help change policies that had long outlived their usefulness. Pictures of the accident at Chernobyl accelerated change in the Kremlin's antiquated news management policies, while images of what appears to be a high-powered laser facility near Dushanbe and an environmental disaster at Kyshtym have generated official Soviet commentary, providing new information for evaluation in the West. More importantly, satellite images of the radar station near Krasnoyarsk may well have contributed to a more activist Soviet stance to resolve this diplomatic embarrassment.[6] Democratic processes in the Soviet Union can still be partially reversed, however. Thus, in the future, there are no guarantees that the release of commercial satellite imagery in the West will elicit positive responses in state-dominated media. In contrast, public access to satellite products are irreversible in the West, providing, in Michael Nacht's words, "a ticket of admission" to participate in a crisis for anyone with an axe to grind.[7] Because of its satellite capabilities, the

Kremlin has an unlimited supply of tickets that it can hand out if it wishes to create difficulties for Western political leaders. The distribution of pictures of sensitive military facilities or operations can be especially discomforting to the West during crises, when they can generate contentious debates or difficulties in relations between America and American allies.

In the past there appears to have been a tacit agreement between the superpowers not to publish high-quality images taken by NTM. In contrast, the United States has occasionally released pictures for public diplomacy purposes of troubling activities by the Soviet Union or its allies taken by cameras aboard aircraft. Commercial observation satellites fall between these two categories. So far, the Kremlin has refrained from releasing satellite pictures in public diplomacy offensives, despite the publication of SPOT pictures of Soviet military facilities in the Pentagon's annual publication, *Soviet Military Power*. Clearly the use of SPOT material in U.S. government publications erodes the tacit superpower agreement not to publicly release images taken by NTM. If the Kremlin chooses to respond in kind to the release of commercial satellite images, it can provide sharper images than SPOT to any nation of its choosing through its marketing agency, Soyuzkarta.

Commercial satellites, therefore, present yet another complicating factor for political leaders and decision makers in the future, whether they operate in relatively closed or open societies. The added power of the media to influence the course of events, especially during crises, should still not be overstated. Any satellite, no matter how extraordinary its capabilities, needs to know where to look to be of maximum utility, and if commercial satellite operators know where to look, the probabilities are that the story in question has already found the front page. Moreover satellite images rarely convey evocative national security messages to the untrained eye. Images taken from aircraft will be more readily understandable, while images taken by cameras on the ground will continue to have the greatest power to activate public opinion. There are few breaking stories or breaking crises that lend themselves to commercial satellite photography. Instead, imagery from space will probably have a marginal role in reinforcing other pressures on decision makers during crisis situations. When it does, a new round of debate will ensue about the influence of the media and how to deal with it.

MULTILATERAL VERIFICATION OR PEACEKEEPING OPERATIONS

A number of commentators, especially abroad, look to satellites not operated by the superpowers as important monitoring tools for future arms reduction and peacekeeping agreements. Other commentators in the United States (with some important exceptions) ally themselves with the U.S. government position, being quite skeptical of these ideas. In the Soviet Union, official views on this subject—like so many others—have changed dramatically, with the Kremlin currently supporting efforts in the United Nations to establish international verification arrangements, including the use of satellites. The advent of new and better satellites operated by third parties during the 1990s will inevitably sharpen domestic debates over the utility of international satellite monitoring arrangements.

The dominant skepticism within expert circles in the United States rests partly on the fuzzy resolution of third-party satellites compared to that of NTM. Nevertheless, it is worth recalling that existing commercial observation satellites offer image quality comparable to America's early photoreconnaissance satellites, which provided a wealth of important information. And over the next decade, the resolution capabilities of third-party satellites will certainly improve, although these platforms might well be too limited in number and in capabilities to serve central monitoring functions for multilateral arms reduction agreements.

Other causes for skepticism about international satellite monitoring arrangements relate to the difficulties in funding a suitable number of satellites and working out operational arrangements for the tasking of such a system and the photointerpretation of its products. Peacekeeping and monitoring functions can be performed by dedicated satellites purchased and operated for such purposes by a nation, group of nations, or an international organization. Most proponents of international satellite monitoring prefer the latter approach, with the establishment of trained cadres of international civil servants to serve as photointerpreters (PIs). Since dedicated verification or peacekeeping satellites would not be continuously engaged in their assigned tasks, they might also be applied to other critical functions, such as environmental monitoring or other uses promoting planetary health and safety.

Alternatively, peacekeeping and monitoring tasks could be undertaken by commercial satellites that can supply nations that are party to an agreement or international organizations with imagery on a contractual basis. If this were the case, sensitive questions of nondiscriminatory access to the photos would have to be resolved because treaty signatories would naturally demand assurance from commercial satellite owners of priority service despite their other business commitments. On some satellite passes, particularly those over Western Europe, this may be a problem because the queue for commercial satellite imagery can be quite long.

Dedicated satellites for peacekeeping and monitoring operations are attractive for many reasons, not the least of which is the avoidance of commercial complications like those described above. They also provide an opportunity, if not an obligation, for all parties to an agreement to participate and share in the cost of implementation. In theory, at least, the more that parties to an agreement have at stake in its implementation, the more interest they are likely to demonstrate in the proper implementation of the agreement by others. Dedicated satellite operations can also soften disparities in NTM between treaty signatories, a long-standing source of concern in multilateral negotiations.

At the same time, there are numerous drawbacks to dedicated peacekeeping satellites. To begin with, these platforms will be quite expensive. A 1988 Swedish Space Corporation study estimated that the cost to build, to launch, and to operate a single satellite is in excess of $400 million, or two-thirds the entire U.N. peacekeeping budget for 1989.[8]

Effective monitoring capabilities from space would probably require multiple satellites carrying different types of sensors, raising costs far higher. The number of satellites under international control would depend on what the signatories of an agreement define as adequate and what they can afford. The requirements need not be too stringent because satellites will likely play no more than a monitoring role supplementary to that of ground-based observers utilizing aircraft and helicopters. Nevertheless, the cost to establish a redundant monitoring capability in space will be substantial. It will be extremely difficult for any international organization to raise funds for a satellite network or for individual states to make significant monetary contributions for this purpose in the face of so many other competing demands on resources. Donations in kind from participating states, therefore, might be a more realistic approach to the establishment of international monitoring capabilities in space.

Other hurdles must be surmounted if dedicated satellites for peacekeeping and verification purposes are to work in practice as well as in theory. The skills acquired for the photointerpretation of verification and peacekeeping functions could also be applied to the planning of offensive military operations, and political considerations reflecting national positions could readily affect the work of an international cadre of trained PIs. The latter is a particularly serious problem because the loss of credibility by an international verification agency as a result of politicized operations would be hard to regain.

Commercial observation satellites have several advantages over dedicated verification platforms if access rights to these images can be properly secured. A modest network of commercial observation platforms will be in place during the 1990s, whereas a constellation of international monitoring satellites is a more remote possibility. The cost of developing these commercial capabilities will be primarily assumed by national governments that might balk at funding dedicated verification satellites. These states might, however, be persuaded to donate commercial satellite services for some verification or peacekeeping operations. If existing national photointerpretation capabilities can be relied on to analyze commercial satellite imagery, the problems described above associated with the creation of an international agency to interpret photographs from space could be circumvented. Instead, the international agency could serve in a far less politicized role as a receiving center and clearinghouse for data.

Notwithstanding these advantages, commercial observation satellites have clear limitations for verification and peacekeeping purposes. As their name implies, they are being operated largely for reasons of commerce, not to promote international goodwill. When tension arises between these two roles, it is reasonable to assume that commercial satellite operators will opt for their primary mission. Other problems may arise when nations financing commercial satellite operations have a stake in a contentious verification or peacekeeping issue. And because these satellites are designed for commercial purposes, their sensors and mode of operation will not be optimal for monitoring purposes. Moreover, the use of commercial satellite images does not address important problems and inequities in the collection and interpretation of images; it merely leaves them at the national level.

As the above discussion clearly suggests, significant problems stand in the way of the effective multinational utilization of pictures from space, whether commercial or dedicated satellites are used for verification or peacekeeping purposes. Nevertheless the quality of these images is improving, and an increasing number of countries are investing in ground stations that permit their rapid receipt and analysis. As this process continues, and as more and better observation satellites become operational, the rudimentary outlines of multinational verification from space will take shape, whether or not nations formally apply them to verification and peacekeeping questions.

The spread of observation satellite capabilities will spark domestic debates in the 1990s about whether the U.S. government should facilitate international monitoring capabilities from space or continue to resist them. Whether the United States likes it or not, the satellites of other nations will be playing an increasing role in international affairs. The challenge for domestic audiences is to move beyond skepticism over these developments and to try to contribute to positive applications. The challenge for those who are highly enthusiastic about new multinational verification and peacekeeping roles for satellites is quite different—the challenge is to move beyond hortatory pronouncements and to consider how the array of operational problems associated with multinational systems can be satisfactorily resolved.

NOTES AND REFERENCES

1. "Should the Senate Ratify the SALT Accords?" *National Review* (July 7, 1972), p. 744.
2. Bernard Brodie, *Strategy in the Missile Age* (Princeton: Princeton University Press, 1959), p. 200.
3. U.S. Arms Control and Disarmament Agency, "Administration Report to the Congress on U.S. Policy on Antisatellite Arms Control, March 31, 1984," *Documents on Disarmament, 1984* (Washington, D.C.: USGPO, 1986), pp. 204–19.
4. See Office of Technology Assessment, *Commercial Newsgathering from Space, A Technical Memorandum* (Washington, D.C.: USGPO, May 1987).

5. This discussion draws on the analysis of Michael Nacht, "Implications for Crisis Decision Making," in Michael Krepon, Peter D. Zimmerman, Leonard S. Spector, and Mary Umberger, eds., *Commercial Observation Satellites and International Security* (New York: St. Martin's Press, 1990).
6. This discussion draws on a more detailed assessment of how observation satellites have affected public diplomacy and compliance questions by Peter D. Zimmerman, "Remote Sensing Satellites, Superpower Relations, and Public Diplomacy," in Krepon et al., *Commercial Observation Satellites and International Security*.
7. Nacht, "Implications for Crisis Decision Making," p. 194.
8. See "Sweden Investigates Satellite Verification of Disarmament Treaties," press release, Ministry of Foreign Affairs, September 13, 1988, p. 3.

12

Security and Technology

John Zysman

Security rests on economic foundations, on the ability to focus as well as to generate resources. It seems evident that in the long run power and wealth depend on each other. The ability to focus resources on security concerns is, evidently, a political matter. The capacity to generate these resources turns on how we structure and organize our market economy. That capacity is a matter of political economy.

I come to the question of security from the vantage of political economy, from the vantage of one who has been looking at the processes of generating wealth and the changes that the processes have brought over the years to the international economy. The inability of American industry to maintain its position in the global market, to adjust competitively to changes in the international economy, is beginning to pose real problems for our security policy.

The security debate since World War II has centered on the Soviet Union, which has been our main adversary. While attending to Soviet developments and challenges, we have often blinded ourselves to the security consequences of the relative change in the industrial position of our allies.

In the last decade a multipolar economic world has emerged. Japan is much more than an industrial and economic challenge. It has become a major player in international financial and trade discussion. In part as a response to the relative decline of America and the emergence of Japan, the European Economic Community (EEC) and elements of business and governmental elites in Europe have launched the movement for a single market. With the announcement that the European Free Trade Association (EFTA) nations will seek to coordinate policy

---

John Zysman • Department of Political Science, University of California, Berkeley, California 94720.

with the EEC, a single European economic market the size of the North American market is coming into being. Whether Europe will develop a distinct security identity and whether that identity will change North Atlantic Treaty Organization (NATO) relationships are matters that may in the end turn less on the capacity to generate resources, which the 1992 story addresses, than on the capacity to focus them and on the objects of focus.

This shift in the relative industrial positions of the elements of the Western alliance is occurring at the same time as the relations between the civilian and military economy are changing. Certainly, military technology must be built on a civilian base, not simply the firms and their know-how, but the foundation of science and technology. However, suddenly we are forced to think again about the notion of spin-off technology, which moves from advanced developments in the military to the commercial sector, and to coin a new phrase, *spin-on technology*, to describe those military systems built from advanced commercial technology.[1]

This chapter proceeds in three steps. The first, of which I am most confident, assesses the changed American position in the global economy and our shift in commercial technology. The second step assesses the changing ties between military and commercial technology, changes that make commercial developments of more importance to security technologies precisely at the moment when America's industrial position is weakening. The third step speculates on the implications of these changes in the international economy for the structure of security relations.

AMERICA'S CHANGING POSITION IN THE GLOBAL ECONOMY

America is not being transformed from an unchallengeable giant into a feeble pygmy. Our capacities have always been more limited than the power we were perceived to have, creating, as Samuel Huntington has remarked, a Lippman Gap between our commitments and our resources.[2] Our economy remains the world's largest; our technological and scientific capacities remain the deepest and broadest.

Nonetheless, while our power measured by trade, money, and technology was once dominant and could serve as a resource for foreign and security policy, our position is now much weaker. Instead of opening our markets to allies in Asia and Europe to cement their ties to us, we now pass trade legislation that, in fact, endangers those ties. At one time we allowed the dollar to serve as the linchpin of the global financial system and resisted efforts to displace it from that role. Now international debt problems require Japanese action. The international financial system depends as much, if not more, on Bonn and Tokyo as on Washington. Instead of simply worrying about the export of American technology to the

Eastern Bloc, we now worry about American access to the Japanese technology required for advanced military systems.

The change in America's position, our relative decline, is not simply a matter of other countries' rebuilding in the postwar years or catching up to our leading position; rather, in the last decade the capacity of American industry to respond to shifts in global markets has deteriorated. That deterioration has, in turn, altered the underlying structure of the global economy. It is a story in two parts.

American Manufacturing's Declining Position

The enormous American trade deficit through most of the 1980s that produced unprecedented international debt is a suggestion of underlying structural changes in the American economy. Underneath the deficit, and the macroeconomic processes that contributed to it, lies an American production problem.

The emergence of the trade deficit doesn't contain many mysteries. The huge U.S. budget deficits financed by foreign funds drove up the dollar's value and priced many American goods out of domestic and international markets in the early 1980s. The mystery lies, rather, in the failure of the trade deficit to be eliminated by the sharp drop in the value of the dollar in the late 1980s. In the 1970s, dollar devaluation had managed to put the current account back in balance. Something was different in the 1980s, and what was different was more than macroeconomic conditions. Simply put, the price elasticity of imports had changed. First, each unit change in the value of the dollar produced a larger increment in imports than it had a decade before. Second, as American producers were driven out of markets altogether or found their competitive position weakened, declines in the value of the dollar didn't reduce our import hunger. Whether under American or foreign control, industries in the United States couldn't product products to compete with the imports.

The Japanese, by contrast, faced a similar situation in the late 1980s. The yen rose in value sharply, leading even comic books to predict factory closures and economic disaster. But disaster didn't come. A trade deficit didn't emerge. Indeed the Japanese trade surplus didn't vanish but stabilized at a very high level. Certainly economic conditions in Japan at the time the yen rose were different than they had been in the United States when the value of the dollar increased. That is not, however, the whole story.

Firms in Japan adjusted to changing conditions, where just a few years earlier their American counterparts did not. Certainly the explanations were multiple, for example, different pricing strategies and different access to the home market. However, an important part of the adjustment came because Japanese firms used manufacturing innovation and advantage to defend their market positions.

The message of the 1980s for the American government may be that the capacity to manufacture is a national asset, and its erosion a national problem. The lesson for companies is equally simple: You can't control what you can't produce.

In history as taught to those of us now in middle age, the United States was supposed to be the dominant manufacturer, and after World War II there was merit to the argument. The United States made things others could not; and products others could make we often made better and cheaper. Our advantage rested on real innovations in the late nineteenth and early twentieth centuries. A system of mass production and divisionalized management underlay our position.

Other countries tried to catch up. They sought to imitate what we did; they saved and invested to do so. In a real sense they never did imitate the United States; rather, they created their own sets of innovations, and from them they built a basis for advantage in global markets.

There were two sets of innovations abroad, one set in policy and one set in production. In a sense priorities are innovations, but in any case Japan and Germany certainly chose to emphasize investment in production over consumption. In both countries the capitalist market economy was organized differently than in the United States. The differences, as we have learned, assisted industrial and technological development, although quite differently in Japan and Germany. After World War II we tended to view deviations from our form of capitalism as either partial modernization or apostasy. We were very slow to recognize that there was more than one form of capitalist market economy. Those differences in policy and institutional structure create distinct patterns of market logic and drive specific patterns of firm strategies. Even now we tend to emphasize issues of fairness and unfairness of other nations' policies, rather than asking how policies shape market logic. Whether specific practices were fair matters less than whether they were effective. Real innovations in production were generated and entrenched.

The second set of innovations lies in production. In countries as diverse as Japan, Italy, and Germany new approaches to production have emerged, providing lower costs, higher quality, and tighter delivery times. It is not an issue of incremental or even radical improvement in an old system, but a new approach— a new paradigm. It is not a story of speeding up the line; rather, it is a story in which the central code word is *flexibility*. In manufacturing, flexibility has two important components: static flexibility, the capacity to vary product mix on a single production line, and dynamic flexibility, the capacity to introduce new production methods and products.

These two analytic components, or capacities, have been concretely combined in two models. One is *flexible automation*, the ability to introduce variety and rapid change into volume production. Until recently, volume production has been dominated by the rigidities of scale economies, expensive equipment dedicated to specific tasks in which the costs could be recouped only by large production runs of the same item. Variety could be very costly; however, through social organization reinforced by technology, variety and rapid change have been introduced. Indeed, the real power of the Japanese system comes from flexible automation.

The other model is *flexible specialization.* Clearly evident in northern Italy and parts of Germany, this model involves an attack by smaller firms on niche markets. It is built on craft skills and an infrastructure of communities that permits horizontal ties amongst firms that compete one day, collaborate the next.

The concepts of flexibility are evocative, and the observed forms of flexible automation and flexible specialization really suggest the emergence of a break from practices dominant in the middle part of this century. The evolution in manufacturing practice can be depicted from another vantage. In a truly remarkable work Ramchandran Jaikumar has depicted the development of the technology of process control.[3] He argues that manufacturing has evolved through six steps, each step involving changes in how people have thought about manufacturing and creating the substantial advances in productivity and quality. Each step has addressed a different source of variance in the production system and, by mastering those sources of variance, has given a burst of competitive power to the newly innovated approach. The first three, I would argue, culminated in the post-World War II American system. Those first three steps were (1) the original emergence of machine tools in England, (2) the establishment of the American system with special-purpose tools and interchangeable parts, and (3) the Taylorist system of people management. Each of the first three steps saw an increase in scale, increasing specification of tasks before production began, and rigid unchanging control of the system once in operation, limiting response to the unexpected inside or outside the production system. The next three steps Jaikumar depicts are, I would suggest, entangled with the Japanese production innovations. Those steps are (4) the introduction of a dynamic adaptive world through statistical process control, (5) the introduction of information processing and numerical control, and (6) the emergence of intelligent-system and computer-integrated manufacturing. These last three steps reverse the trend toward scale and of tightly managed control of people in the production process. The system built in these last three steps is adaptive, with extraordinary levels of productivity and quality. Jaikumar estimates that the fully developed six-step system would have a productivity increase of between ten and fifteen to one over the three-step system of the American years. Using product rework as a measure of quality, quality is improved by two orders of magnitude. In my view we are now effectively moving through the fifth phase, with the sixth phase clearly imaginable but perhaps still somewhat beyond our technical and organizational capacity. That implies, however, that those who fully implement the fifth stage today will generate productivity of five times, and quality an order of magnitude, better than that seen in the fully developed American model. That potential is consistent with what has been observed in a range of industrial sectors over the last years. Radical discontinuous jumps in production technology have created distinct competitive advantage.

Importantly these production innovations do not stand alone. It is not simply that cost is reduced, though it is, or that quality is improved, though quality is an inherent product of the innovations. Rather, rapid introduction of new products—dynamic flexibility—and rapid variation of established products is also facilitated. That package of advantages becomes a powerful competitive weapon allowing a whole variety of new marketing strategies to translate production advance into market position.

The question is why this second set of innovations occurred in Europe and Japan and not in the United States. The first answer is simple and obvious. A dominant and effective system existed in the United States. Until that system was challenged, there would be little need or incentive to alter existing practice. However, that still leaves open the question of why the challenge emerged as it did. The answer is that the logic of firm choice in Japan and parts of Europe generated that response. Consider the Japanese case. After the war Japan's market was protected and growing rapidly. Its firms were technology followers borrowing technology abroad. Firms faced the need to borrow and rapidly implement technology from abroad. Each market increment that came through growth allowed the possibility of borrowing and implementing another round of technology. Thus, in quite traditional industries, Japanese firms faced conditions that Americans associate with high-technology industries. Learning-curve economies dominated, making the pursuit of market share a necessity to sustain short-term profits. In that environment and with capital short, a system emerged of semi-market ties between assemblers and component producers that organized production in new ways. The organizational innovations, including the use of statistical process control, moved firms into the fourth of Jaikumar's stages. That the technology was shaped by the new choices, rather than driving them, can be seen in differences in American and Japanese machine tools. What seems common to the Japanese, German, and Italian cases are broad definitions of job responsibility and highly skilled and educated work forces.

The problem for America lies not in the pressure from abroad, which is advancing production and product development to our advantage as well; rather, the difficulty is in our response. And our response has been driven by mythologies that have been very difficult to shake off, mythologies that have affected both government and firms. In the policy arena the notion of sunrise and sunset industries distorted our understanding both of trade and of the economy. We were slow to grasp that the bulk of the sunrise industries produce intermediary goods used in the products and production of other industries. Consequently the so-called sunset industries were the clients of the so-called sunrise industries.

The problem was how to use the new transformative technologies to alter traditional industries. Similarly the notion of a postindustrial society kept us from understanding that we were witnessing a transformation in industrial production, a shift in the role of services in manufacturing, not a move up and out of industry.

Similarly there were a set of corporate myths that are worth mentioning if not developing here. The first was the notion that one could win with technology, leaving the dirty business of production to others. Certainly this led firms to cede parts of the market where production mattered most to foreign companies that then built distribution channels from which to attack the technology-intensive segments of the business. Equally, it deceived firms about the nature of product innovation. Product and process knowledge are not that separable, and except for a few disjunctures, new products are built from knowledge accumulated in early generations. Cede production and you limit product innovation. A second myth was the notion that our competitors' advantage lay with cheap labor. That hid from view the powerful evolutionary steps in manufacturing going on abroad. It led to moves offshore that established networks of production that made Asian production cost-effective, quite apart from direct labor costs. A third myth was the notion that capital costs alone kept American firms from an effective use of technology, when—as the General Motors case reveals clearly—the central obstacle was an understanding of what to do.

In sum, what has been lost? Oversimplified, it is not simple market position in specific sectors. Nor is it simply the ability to produce in specific sectors. Rather, we have lost in some sectors and are losing in others the capacity to sustain competitive production. This is reflected in many production equipment sectors, from machine tools and robots through such semiconductor devices as photolithography equipment. It is reflected in the difficulty of obtaining competitively many components in the United States, and most seriously in the ability to find the skilled workers and designers needed for reentry into domestic production. Then, of course, the ability to design new products in the future rests on the knowledge and the profits of an existing market position. The weakening of American industry in global markets is serious and will make the recapture of future competitiveness more difficult in the future.

A Multipolar Global Economy

The upshot of the eroding position of the United States in industrial competition has been the emergence of a multipolar global economy. Even as the markets for goods and money became international, limiting national control, regional economic groupings became more significant. This is not simply a matter of postwar reconstruction's ending the artificial absolute dominance that America held after the war. It is not simply a matter of the decline of a hegemon; rather, it is the evolution, since the oil crisis, of new, regionally based production capacities and two regional powers that are now allies and competitors but could in the future be rivals to the United States. Simply to examine gross national product (GNP) statistics that note the continued importance of the United States in the global economy would be very misleading. Just as simple reference to manufacturing employment or production hid from us the profound changes in global manufacturing, so simple GNP numbers hide the importance of these regional blocs. Let us look, in turn, at Asia and Europe.

The Asian economic region has Japan at the core and a series of vibrant, expanding economies emerging around it. The regional strengths lie partly in production advantages by firms of the sort noted above. While the production advantage in Japan rests increasingly on production innovation, elsewhere it continues to rest with more traditional approaches and lower labor costs. However, it is not a matter of lower direct labor costs, which are increasingly small portions of production, but of highly skilled and often highly trained development and support staffs that are also less expensive than in the United States.

Asian production strength does not rest only on the organization of the individual firm. Rather, production advantages lie in the networks of component and product companies that have appeared in the last decade. As American firms moved abroad seeking low-cost labor, they transferred product and production technology to the region. In some cases the American firms organized production themselves, training workers and managers. In other cases they subcontracted production. In sum, American firms helped create a whole infrastructure of firms in Asia that together, now a generation later, create cost, delivery, and quality advantages for our competitors.

Increasingly the Asian complex generates product advantage as well. Certainly Japan is the center of this development. In consumer durables—electronics and automobiles being the clearest cases—there are many products that the United States (and often Europe) simply cannot produce competitively. Korea has taken a competitive position in some goods based on low labor cost, scale, and established designs. Taiwan often builds design advantage and not simply cost advantage. The automated production facilities of Wyse Technologies is a good instance. However, consumer durables are not the only examples.

Importantly, Japanese technology and components are at the core of the network. Korean televisions and cars, for example, depend on Japanese components. Thus a production core independent of American technology and know-how, though tied to American markets, has emerged. As incomes rise, an Asian market may emerge, further disconnecting the Asian economies from the United States.

Japanese growth continues at a pace that exceeds that of the United States and Europe and that will steadily increase Japan's relative power. There are critical continuities in Japan. Where once apparent weakness and market closure created the basis for low-cost technology transfer, now financial muscle and market strength continue a flow of foreign technology toward Japan. The asymmetry of market access—whatever the mix of causes among policy, business practice, and client finickiness—continues as a strategic advantage. High saving rates will continue to fund domestic development.

The emergence of a political core in this regional economy may seem doubtful, but there are dramatic developments. A new Japanese power is beginning to find expression. Japan, not the United States, is now the leading foreign aid provider. Similarly Japan and not the United States has the resources to structure the resolution of global debt problems and, in alliance with Germany, perhaps even to structure the global financial system. The enormous political advantages of being a global reserve currency are understood in Japan, and one can argue that the yen will slowly assume at least some of those functions. American acquiescence to an increased role for Japan in the International Monetary Fund is an expression of that changed economic power. An unraveling of the dominant Liberal Democratic Party may slow this process. Finally, and we will return to this matter, Japanese commercial technology has now established a basis for independent choices about military force composition.

At the same time, the emergence of a European economic region with some significant political coherence now seems a likelihood. The Europe 92 movement is a response to the emergence of Asia and the real decline of the United States as a source of technology and production know-how. For the last two generations Europe's economic position has rested on a set of implicit bargains with the United States. Europe had access to American technology; even as it trailed in the development of advanced technologies, it excelled at applying them. Its position of privileged second might have been grating, but it was tolerable and did not provoke joint European action. Suddenly, however, crucial technologies appear to be available only from Japan. In finance the dollar anchored the international financial system. This provided privileges to the United States but gave others stability and (at least until 1971) the right to devalue against the dollar to maintain trade equilibrium. Now Tokyo and Bonn as much as Washington could dictate financial evolution. In trade the American market was open while the United States accepted and encouraged the creation of the European Community. Recent American trade legislation now threatens to close our domestic market, or at least raises that possibility, while the Japanese market is relatively impermeable to Europe and the United States.

The implicit economic bargains between Europe and the United States were set inside of explicit security bargains. Put aside arguments about culture or history. America and Europe share a security problem, but Europe and Japan do not. Consequently relative dependence on Japan in finance and technology and asymmetrical market access make it unattractive for Europe to exchange America for Japan as hegemon.

Europe's technology position is changing significantly. Europe's fundamental strengths have always been underestimated. They rest on an educated and highly skilled work force, a sound foundation in science, and the enormous wealth built up through a long and successful industrialization. New strengths have been added to this older foundation. Those strengths lie in the application of advanced technology to traditional industries, a capacity for systems development and integration, and the use of political will to retain final product markets in the face of production or product advantage. The most obvious weaknesses of the postwar years are now being overcome. For example, the commitments in microelectronics and telecommunications made by national governments, the community, and individual companies are beginning to succeed. Siemens and Phillips are building a real capacity in dynamic RAM, and Thomson is emerging as a serious player in the global semiconductor market.

In sum, the Western economy increasingly consists of three economic regions. Whether those become sealed rivals will depend on the politics of the next few years. More has happened than the end of economic hegemony. It is not merely that a relatively weaker America stands among a set of relatively stronger middle-sized economies. The United States is increasingly dependent on technology and finance from abroad, dependent on two security allies who have increasingly become commercial rivals; the Western economy has become multipolar.

FROM SPIN-OFF TO SPIN-ON TECHNOLOGY

The links between military and commercial technology are changing. For years the military helped justify expenditures on the ground that investment in advanced technologies first applied in military or aerospace uses would spin off into commercial applications. Extensive studies, in fact, attempted to quantify the process or consider how it could be accelerated. Now, suddenly, from Japan we find the importance of spin-on technologies. The latter are technologies that have become established in the commercial sector, but which are directly or with minor modification the basis of advanced military systems. Commercial technology, always the foundation of military technology, is of increasingly direct importance to security. That it has become so precisely at the moment when America's industrial position is weakening makes the issue all the more important.

The model in the years after World War II was that investments in big science and advanced technology would have their first applications in the military sector. Only the military could support the enormous development costs. They could—indeed needed—to do so, because early use of advanced technology would create military advantage. Thus, in jet aircraft, semiconductors, and computers, the military was seen to have played a vital role in developing a new industry.

The military role, however, was not always what it seemed. In the case of the semiconductor industry, for example, it was not government development expenditures that accelerated the use of technology. Indeed government investments were often misplaced. Rather, the government as first user ensured an early launch market. Loan guarantees for production development helped firms play that role. Of course, in aircraft the military development of jet bombers certainly underwrote the commercial industry. Indeed that most profitable of planes, the Boeing 747, was never sold to the military; but Boeing initially made the investment for the transport in anticipation of an early military market.

Science and advanced technology, it seemed, had changed the relation of military and commercial technology after World War II. During the war the fundamental production capacity, the ability to produce tanks and planes in great numbers, proved crucial. There were vital technology developments, including radar and artificial rubber. However, the defense production base was the commercial base. The relations between the military and civilian sectors alter over time; now one is the leader, now the other. Certainly the early development of mass production and interchangeable parts was accelerated in the United States by the military demand for rifles in the Civil War. Since World War II, however, the new science-based technologies appeared to separate the problem of advanced weapons development from that of traditional commercial products. Its role of technology driver seemed to make the military an instrument, not only of military development, but of commercial advance.

Now once again the relationship between military and commercial technology have shifted. Advanced military technology is increasingly built upon a commercial base, as was the case in World War II. This is not, however, a return to an old model. Advanced military systems still involve new science-based technologies and the development of unique systems from these new technologies. But the cutting edge of new technology is now increasingly driven by commercial—not military—markets.

Let us note two such cases. Heads-up displays and location technology, which we might associate with advanced fighters, will become a standard of the automobile industry. Chips that control an auto engine or braking system are in an environment that is often as hostile as the battlefield. The advanced microelectronic circuitry of a digital television set will be of the same sophistication as advanced computers; the display systems for television will push the edge of display for the most sophisticated scientific uses.

Several processes are at work here. First, the basic technological requirements of new consumer products now approach or equal those needed for more sophisticated applications. Second, the enormous volumes involved in consumer durables makes the commercial market even more attractive than military markets. Third, because products require advanced technologies for the development of which companies will pay out of their enormous profits from commercial markets, purely commercial producers can support research and development

(R&D) efforts associated with military applications. Fourth, for commercial consumer applications the unit costs of the component technology have to be very low. Yet a real-time processor for engine or brake control on an automobile is a very sophisticated element. Low costs cannot be achieved by reduced functionality. The enormous volumes allow auto producers to negotiate with semiconductor manufacturers unit costs an order of magnitude lower than those projected for scientific or military applications. Fifth, volume production also involves new manufacturing approaches and technologies for advanced products. Industries, such as the computer industry, which have been involved in batch production suddenly find themselves in mass or process production settings.

The speed of product development and new component introduction is so critical that it should be emphasized. The commercial pace of application of advanced technology to product and system is accelerating. Market competition is forcing the pace, and reorganization is allowing firms to respond. Honda can now take an automobile from design to showroom in less than three and one-half years, twice as fast as can Mercedes. With the increasing application of electronics and materials to automobiles, the problems are similar to those of military development.

The American military has recently been concerned that the time for military systems development has become so long that the components used in the system are often two generations old. The component technologies are advanced as design begins but obsolete when production starts. Consequently programs such as Very High Speed Integrated Circuits (VHSICs) are launched not only to advance the technology but to link systems and component developers. Can civilian developers move complex systems from design to battlefield faster than traditional military suppliers? Is the fastest route to the most advanced systems through two complete generations of product made possible and financially feasible by new design and production organization and technology? There is every reason to believe that spin-on technology will include the basics of product design and development.

This changed relationship may mean that Department of Defense (DoD) initiatives may no longer prove effective means of spurring either technology or commercial position. The VHSIC program, for example, seems to have proved largely irrelevant to the evolution of the semiconductor industry. More strongly, DoD initiatives, at least as currently structured, may serve to impede development. Organizational structure in firms dependent on the military is shaped by the dominant client. In this case the organizational structure, product, and production preferences of military contractors seem to have been structured by military programs in ways that reduce their effectiveness for commercial objectives.

Technology's evolution follows trajectories that reflect the community and context in which it is developed and that are not inherent in the technical knowledge. Technology is a path-dependent process of learning in which opportunities for tomorrow grow out of research, development, and production undertaken today. Massive resources committed to specialized defense contractors for technology produced in batch processes for initial use in military projects will constitute one trajectory. Massive resources committed to commercial development produced in volume for consumer markets will constitute a separate trajectory.

Civilian and military initiatives may represent two different ways of developing advanced technology. The U.S. Air Force supported the development of advanced machine-tool technology for application to advanced aircraft. The programming language proved too complex for commercial applications. The Japanese Ministry of International Trade and Industry supported the development of such machine tools, in some versions, forcing a single controlling supplier to allow competition around commercial application and, in other versions, encouraging diffusion. In either case, commercial applications drove the Japanese industry. The resulting machine tools were lighter and simpler.

The difficulty lies in the fact that these two trajectories may represent competing, not complementary, routes. Power and wealth in the long run depend on each other. In the short term, however, states frequently have to choose between them. Advanced military and civilian technologies increasingly converge on the same key sectors. The overlap encompasses the whole technological base of the nation: its research institutions, its industries, its pool of scientific and technical personnel, and its R&D investment pattern—in short, the entire framework supporting innovation. The dilemma is that military objectives for technological development may shape the technical base in ways that handicap the corresponding civilian industries, or so that commercial R&D will not produce the applications.

For the United States the shift from spin-off to spin-on and the potential conflict between commercial and military trajectories pose policy problems. Are our approaches to military development obsolete for their own purposes? Are they counterproductive for the long-run development of the national industrial base on which they must rest?

WILL AMERICAN INDUSTRIAL DECLINE RESHAPE THE SECURITY STRUCTURE?

Will the emergence of a multipolar global economy produce a multipolar security structure? Let us readdress this most basic matter by asking (in three separate domains) how reduced American industrial and economic strength might influence our position.

The Economic Projection of Influence

Reduced economic resources almost certainly limit the capacity to project national influence. It has become a convention of discussion that the United States has consistently used its surplus economic resources for diplomatic purposes. Surplus resources could be defined as the ability to give up domestic production or consumption for foreign-policy purposes without intense domestic political opposition. These surpluses provided us with the capacity to establish and to control institutions in the global economy, and that capacity helped define us as a hegemon. These surplus economic resources meant much more than money for defense expenditure obtained through tax or inflation; they involved choices about trade and financial policy that structured options for business here and abroad. Market access was provided for Japan and the newly industrialized Asian countries of security significance. Surplus resources were also evident in the use of foreign aid for political purposes.

How to measure our reduced surplus resources may be debated. It is not so much a matter of weakness, of a pygmy America, but rather of reduced supplies of surplus resources. And it seems unquestionable that the supply is reduced. The range of instruments for using economic strength as a foreign political weapon is restricted and their effectiveness reduced. The dollar's break from gold in 1971 ended our ability and willingness to maintain the postwar fixed-exchange-rate system, but did not end our position as the dominant financial power. Our inability to absorb Latin American exports to finance their debt reflected a real and intense resistance by American industrialists to ceding market share to bail out the banks and our Latin trading partners. Our trade deficit and foreign debt begin to force us to adapt our domestic policy to international constraints rather than—as before—using domestic policy as an instrument of foreign influence. Our inability to ensure the implicit postwar bargains made to our European allies almost certainly provoked the movement toward 1992. If force is of reduced utility in a nuclear world with regional politics shaped by mass mobilization, economic influence may be of heightened significance. Yet the American capacity to exercise such an influence is now increasingly constrained.

Security and Military Equipment

A reduced economic position, more particularly a declining industrial position, will have its greatest influence on the military equipment we develop and procure. While this matter cannot be addressed apart from the evolving security problems we face, there are significant developments that are not simply a matter of how much money we spend. The crucial issues may be the technology available and the development and production costs of systems.

The United States must confront the question of whether dependence on foreign, though allied, sources for crucial military technologies is a concern. We may decide that such dependence is acceptable, but we should not delude ourselves into believing that this dependence does not represent a change or that it does not matter. Our control over vital technology has in the past provided us with a lever in discussions with allies, although when made fully explicit that lever has often been counterproductive. There is no reason to believe that others will not use similar leverage that they may have to influence the course of discussions with us.

How will diminishing dependence on American technology affect interallied relationships? The real lesson of the FSX fighter is not about technology transfer, but about Japanese ability to assemble their commercial technologies into military packages. Many observers believe that Japanese-developed avionics will be superior to those developed in the United States.

A broad range of technologies now exists where the United States is dependent or risks dependence on foreign sources. The debate about microelectronics has focused attention on Japan. However, new military telecommunications exchanges deployed since Grenada are of French technology. Recent procurement for critical radar display technologies saw no American bidders. New materials may be crucial for *Stealth* systems, but they are equally an element in next-generation automobile production and a focus of commercial attention in Japan. The list of technologies will grow.

Every bit as important as component and systems technology is production technology. Assume a national defense production system that employs next-generation production technology in which the costs of small-batch production approach the costs of volume production, or at least a system in which the penalties of batch production are reduced. In such a system, a given procurement budget will buy more weaponry. A middle-sized power with a limited defense budget could extend its force projection capability. It would be revealing to look at current and planned production costs in Japanese tank and ship programs to explore how existing commercially derived production systems affect weapons development.

Assume also that our next-generation national defense production system attempts to take advantage of the speed from design to deployment implicit in commercial markets. For our purposes assume that the development speed is accelerated so that two product cycles can be completed in the time a rival completes one, as is the case, for example, for Honda and Mercedes. We will even grant that each weapons cycle stops short of attempting the most sophisticated system possible in the interests of speed and costs; but the speed to the next system makes that trade-off less painful. In such a case our existing assumptions about component obsolescence and system reliability would no longer be valid. Deployed weapons would use components that are closer to, if not at, the state of the art. Since slightly less radical jumps are attempted in each cycle than is now the case, the systems, furthermore, would be more reliable. The logic is simple; an advanced production and development system could provide reliable weapons systems at a reduced cost using state-of-the-art components.

Despite these developments Americans may still assume that they now dominate the state of the art in military systems. While that domination will certainly continue for the next few years, it is by no means ensured in the future. If such a system of commercial production technology provides the platform for military production, then many of our premises about force development may need to be reexamined. Not only can more limited resources produce a given set of weapons or weapons systems, but the state of the art may be pushed ahead more rapidly or in new directions. It is at such moments—when new technologies, production systems, and organizations are introduced—that the dominant producers in commercial sectors lose relative position.

Will New Players Alter the Security Configuration?

The Western economy has become a multipolar system increasingly organized around three defined regional economic groupings. Both Japan and Europe have the resources and the technology to establish independent military positions. Each is steadily less dependent on American military as well as commercial technology. The FSX fighter deal, we should not forget, was itself a compromise to dissuade the Japanese from building an entirely independent fighter plane. Japan's commercial technology has established a base on which Japan could, should it wish, develop its own military system, whatever American wishes might be.

The question becomes a political one regarding European and Japanese purposes, not one of relative technological or economic development. For Europe the issue is whether a common purpose can be defined or pursued. For Japan it is a matter of whether it will choose to assert a more extended military position.

We risk fooling ourselves very seriously. It is probable that as long as American, European, and Japanese interests do not diverge too sharply, neither Europe or Japan is likely to assert a stronger defense position or challenge the Western alliance structure with the United States at its center. Yet a situation in which the United States continues as the alliance leader because our allies can't define an alternative or choose not to do so is radically different from one in which our leadership is produced by our strength. At an extreme, if American military expenditures are financed directly or indirectly by loans and investments offsetting our trade deficits, the United States provides the mercenaries for an alliance system in exchange for increased American debt and control of our assets. This is not a desirable bargain.

The more immediate question is under what circumstances American interests and those of Europe and Japan might diverge sufficiently to provoke either to adopt a more independent stance. There seem to be two sets of such circumstances that we can envision. First, there might be changes in the strategic context or, importantly, perceptions of that context. Europeans will probably interpret and react to changes in Eastern Europe differently than Americans. Evolution in Eastern Europe is likely to determine the course of the 1992 movement in Western Europe. Alain Minc has argued[4] that Western Europe was created by the Iron Curtain and would tend toward a different axis once the symbol of the Berlin Wall was dismantled. Conversely, if Mikhail Gorbachev were to fall and a radical conservative or military regime to take power, Western Europe would, in my view, respond with a much more unified defense posture. Japan's strategic position and problems are quite different. One might speculate on a range of developments in Asia that might lead Japan to conclude that it requires a broader range of policy choices. The second set of circumstances turns on American choices that might provoke European or Japanese responses. An American insistence that Europe and Japan pay for their own defense generally translates into a notion that they pay the United States to do it for them. However, substantially increased defense expenditures in either Europe or Japan will mean independent development of weapons systems. Radical cutbacks in American expenditures or an apparent unwillingness or inability of the United States to provide for the common defense would likely mean a similar outcome. Weapons systems such as the Strategic Defense Initiative, which reduce the equivalence of San Francisco and Rome in strategic reasoning, could have similar consequences. Simply put, we cannot continue to assume that Japan and Europe will refuse indefinitely not to act on their national capacities and interests, and not to become stronger and more independent players.

Europe and Japan, however, will remain American allies, even if our interests and objectives in specific settings and issues diverge. Does this fundamentally change the bipolar structure that has characterized the post-World War II international system? Without entering a debate about what creates alliances and competing alliances, it does not seem likely that the Western alliance will splinter into competing blocs. Rather, we seem to be moving toward a new structure of the international economy involving a continuing bipolar competition, albeit with collective leadership in the West and a sharply diminished capacity for the United States to define purposes or compel action. It will be a difficult, even dangerous, world and a troubled transition.

NOTES AND REFERENCES

1. Steven K. Vogel, *Japanese High Technology, Politics, and Power* (Berkeley BRIE Research Paper no. 2, March 1989).

2. Samuel Huntington, "Coping with the Lippman Gap," *Foreign Policy* 66 (1987–1988), pp. 453–77.
3. Ramchandran Jaikumar, *From Filing and Fitting to Flexible Manufacturing: A Study in the Evolution of Process Control* (Cambridge: Harvard Business School, Working Paper, 1988).
4. Alain Minc, *La Grande Illusion* (Paris: B. Grasset, 1989).

13

Predicting the Future of American Commitments

George H. Quester

One question Americans have asked themselves since the end of World War II is whether they will stay committed to the defense of the democratic nations of the world. In the face of possible menaces from regimes like the Soviet Union, will they remain engaged, militarily and otherwise, continuing to risk conventional and nuclear war, and continuing to invest resources in the preparation for such wars? Another question is whether the commitment will continue in light of the proliferation of advanced weapons.

The repeated raising of the first question reflects an unnaturalness Americans still feel about overseas commitments, and indeed about any active participation in international relations. It also reflects their memory of the rapid withdrawal of the United States from the world at the end of World War I, a withdrawal involving the rejection of the very League of Nations that had been designed by the American administration under Woodrow Wilson.

The question has often enough been accompanied by dire predictions that the United States would soon enough retreat into isolation of one form or another, as Americans would get fed up with the Western Europeans, who have been selfishly letting the United States carry the burden of European defense, or as Americans would get disillusioned about Southeast Asians or Central Americans, who only pretend to be committed to democracy but actually are no more committed to free elections or honest administration than are the Communist factions the United States has been opposing.

Yet, if such predictions are being widely circulated now at the beginning of

George H. Quester • Department of Government and Politics, University of Maryland, College Park, Maryland 20742.

the 1990s, we must note how often they have been put forward (wrongly) over the past four decades. It may be quite fashionable, or an easy way to capture attention, to predict that the relationships of the United States to its allies cannot persist, but the forces for continuity since 1945 have been surprisingly powerful.

This basic question can serve as a natural vehicle for an analyst trying to sort out what makes the United States or any other nation become active in its foreign policy. Is it something very benign and liberal about the United States, whereby it feels a natural obligation to share with others, as much as it can, the liberties and freedoms and institutions that have basically made Americans happy with their own political system? Or is it some kind of malign thrust of capitalism, as the economic system makes Americans pursue markets and economic outlets abroad (i.e., instead of being an unusually good country, would the United States turn out to be an unusually bad country)? Or is it that the United States is merely an ordinary country, power oriented like all the rest, seeking military and other power advantages wherever they become available, pursuing the material and other advantages such power can bring, or acting in fear of what someone else might do with the same accretion of power?

A different form of the basic question about the American international role can also be wrapped around this basic issue of the continuation of the American commitment. Has the period since the end of World War II somehow been abnormal because power politics became bipolar rather than the more normal multipolar, or because Stalin was an unusually villainous dictator (with his successors being more typical statesmen), or because the sheer novelty of nuclear weapons made the times abnormal—demanding a greater American role until everyone could get used to the novel characteristics and magnitude of such weapons? Or is an active American role in world politics something that should be seen now, and should also have been seen in the past, as quite normal, with all of the day-to-day burdens of military and foreign policy being something that it was impossible for the United States to escape? If nuclear weapons have indeed changed things, moreover, they are perhaps now going to be normal for as far as we can see into the future, with no relief or substitute in sight for what has been viewed as mutual deterrence or mutual assured destruction.

Change has more often been predicted than it has occurred. Yet change is always possible. Americans may change in what they feel about the world, and they may change in what they see (correctly or incorrectly) as possible in the world. These are what an economist might call the *indifference curve* and the *opportunity curve* that Americans will face.

First, we will attempt to sort out some trends in the nature of American values themselves, in the principles Americans will espouse and in the human identifications they will feel. Then the discussion will shift to developments in physical capabilities abroad, developments that may alter American attitudes on defense-policy arrangements, with many of these being those changes in military situations discussed in more detail by other authors in this volume.

WHY AMERICANS CARE

The following is proposed as a plausible list of why Americans care about the defense and security of any particular place in the world. The list encompasses some of the alternative models of what explains a foreign policy (and of what is normal or abnormal in such a foreign policy) cited above. From the possibilities on this list we would then have to consider how the future is likely to change Americans.

First, Americans might be concerned about any part of the world simply because of the resources in the area—the resources necessary to national defense or required for material well-being. It is difficult to grow bananas or coffee inside the United States. If anyone were to conquer and cut off the sources of these key ingredients to a morning breakfast, many Americans would feel that they had suffered a major blow to their lifestyle—and even more so if someone were to cut off access to the petroleum resources of the Persian Gulf or to the key minerals produced in southern Africa, some of which are crucial to the production of the latest weapons systems.

Some analysts would argue that the industrial and military potential of Western Europe and Japan is essential to the security of the United States (even if this is mostly held in reserve, because these nations do not devote as large a share of their economic potential to security uses as does the United States). Whether this is the only explanation, or even the primary explanation for why the United States is committed to the defense of Japan or Western Europe is something we shall have to debate. The argument that such considerations are primary would apply much more plausibly to the Persian Gulf and its oil, which Japan and Western Europe so badly need and the United States may indirectly need.

Second, one can be selfishly concerned about maintaining a buffer between one's own homeland and the nearest hostile military forces. It is regarded as generally better to defend yourself in someone else's backyard than in your own. The collateral damage that is inevitable in such a war falls on the territory of someone else, and the chances of your being totally defeated are reduced because you have greater maneuver space and earlier warning of the threat.

The United States would want to have forward bases for its navy, air force, and army because such bases can substantially enhance the military power available in a confrontation with the Soviet Union or with any other enemy. Necessarily, and almost inevitably, the United States will similarly want to deny the use of such base areas to the forces of the Soviet Union because any projection of Soviet military and political power onto the continents of the world could accumulate to the disadvantage of the U.S. national power position.

Third, also still cast in largely selfish and power-oriented terms, the United States can become necessarily intent on defending an area merely because it has defended such an area successfully in the past (i.e., because a precedent has been set), and failure to hold the line now would be seen all around the world as a sign of weakness. Because Americans defended Korea in 1950, they may still have to defend it now; and because they held on to West Berlin by an airlift in 1948, they have had to stand up for the freedom of this once difficult-to-defend enclave. Similarly, because the United States successfully opposed the deployment of Soviet missiles to Cuba in 1962, it would have to oppose the introduction of such missiles into that island at any time in the foreseeable future.

How the world sees U.S. willingness to stick by the precedents it has established plays a very important role in how much power and influence the United States actually has because there are many powers that will rally around a winner but will abandon a state when it seems to be declining in resolve and power.

West Berlin may be the perfect example of a position Americans had to defend, not because it offered key resources (it actually was a drain on the resources of West Germany and of the West in general) or because it offered a valuable military base (the base was far too vulnerable), but simply because of the precedents and images involved. The Falkland Islands may have been the same for Britain.

It should be noted that it is very difficult to get off the hook of such burdens and obligations of precedent once they have been established. World wars have in the past served to wipe the slate clean—no one after 1945 had to be as concerned about Danzig as before 1939—but we all pray that there will be no such cataclysmic wars now to erase our memories. Disengaging from such an implicit commitment may otherwise be very difficult, and disengaging may also be undesirable if one feels that there are other reasons to defend the area in question. (One could imagine designing some face-saving way for Britain to give up the Falklands, with all 1,800 people there being resettled elsewhere with an enormous cash compensation, and one could perhaps even imagine a similar resettlement of all the West Berliners. Yet how to do this gracefully, without stirring up images of weakness, without encouraging aggressors to challenge one's national power at many other points around the globe, is never so easy to outline.)

Finally, and most important, one has to consider the possibility that the United States is committed to the defense of so many nations around the world not because Americans care about resources or want a buffer, or even because of precedent, but because they care about the people who live in these territories for their own sake; that is, Americans identify with them. (All our examples may be complicated and intertwined in the motivations they illustrate. If Americans did not identify with the West Berliners at the beginning of the Berlin Blockade, they certainly had come to identify with them by the end of that crisis, even while the issue of precedent then also remains powerful.)

As a relatively pure example of the power of identification, one might cite the case of Israel. That country does not offer the United States scarce and valuable resources; indeed the Americans' commitments to Israel have in the past threatened their access to the oil of the Arab countries. Israel similarly is not to be prized as a buffer or a military base because the United States might have had a better array of bases all across the Middle East if it had been free to form full-fledged alliances with those Arab states that (out of Islamic fervor) have been concerned about atheistic Communism. Because the United States was not so clearly committed to the defense of Israel at earlier stages, the issue of precedent is not nearly as strong as it is in the case of Korea or Berlin.

Rather, the basis of U.S. commitment to Israel lies in the admiration of and identification with the people, ties of kinship and commitment to similar philosophies, and Israel's position as the only democracy in the Middle East.

ETHNIC CONSIDERATIONS

For any country's foreign policies, not just for those of the United States, predicting such commitments to people for their own sake is not easy. One will find a somewhat uneven pattern of identification with other peoples abroad, a function of similarities or differences in language and culture, in religion and lifestyle, and in simple physical appearance. One shares more vicarious pleasure (and more vicarious suffering) with the people with whom one identifies than with people who are different, and it would be to change the most basic facts of psychology to eliminate this kind of discrimination.

In light of these considerations it is not surprising that Americans have cared about the future of Europe more than about other places. It is from Europe that most of the ancestors of Americans and the languages that they speak have come. Europe is where American political and cultural systems developed; it is a region that has developed along some basically similar (American) paths since World War II (although Europe also took some horrifyingly different paths up to and during that war).

The links with Europe were thus established by the original patterns of transatlantic migration and by subsequent waves of renewing migrations. Until very recently the bulk of new arrivals in the United States have come each year from the European continent, for a time because the immigration laws of the United States themselves discriminated in this direction, afterward because of more natural patterns.

Yet the most recent waves of migration, legal and illegal, now suggest the possibility of a different orientation, one toward Latin America or toward the Pacific rather than the Atlantic, as great numbers of Mexicans and Central Americans, and Koreans and Chinese and Vietnamese have come to the United States. These new Americans have retraced many of the steps of earlier immigrants in the process of assimilation but are not necessarily repeating all the earlier patterns, and they are in any event oriented toward sources of origin other than Western or Eastern Europe.

Might it be that this change of source of population will reduce American interests in shielding Western Europe against a Soviet attack? It is often argued that the United States will see itself now as a Pacific Rim country rather than as part of the Atlantic alliance, with changes in trade relationships also somehow being seen as a major factor in this shift. Such forecasts may amount only to one more wave of the fashionable predictions of breakups and crises in alliance relationships.

The source of language will not change unless the great number of Latin American immigrants somehow succeeds in making Spanish as viable within the United States as French is in Canada. The geographical origin of the Magna Carta and most of the political and philosophical ideas about which Americans argue will also not change because the United States remains a remarkably European country in culture.

To have all of Europe subjected to dictatorial rule will for most Americans remain just about as much of a personal disaster as it would have been in 1940, and as yet we do not have an indicator of a major change in American alliance and defense commitments in Europe. A significant part of American foreign-policy energy is indeed very altruistic at its base, identifying with other peoples for their own sake.

Misery and famine and dictatorship and war anywhere will grip the imaginations of Americans and motivate them to try to do what they can to prevent and alleviate those crises. The important point is that Europe has got and will continue to get, a larger-than-average share of this kind of altruistic identification. Only the most cynical observer would dismiss all this to argue that a nation like the United States is motivated only by personal advantage and calculations of power.

Apart from changes in the immigration patterns, one must note some other trends in the background that determine American identification or nonidentification with peoples abroad. The United States has felt close to the Chinese people for most of this century, but for two decades after the Communist takeover in 1949 it felt very much rejected by the bulk of the Chinese, maintaining its identification only by the pretense that the Republic of China on Taiwan was the proper government of all the Chinese. The wrenches of the Great Cultural Revolution, and then of the subsequent de-Maoization of China, suggest that some great and surprising changes are still possible in China, with consequent possibilities of marked changes in American attitudes toward things Chinese.

Revolutionary turmoil in Iran similarly changed American attitudes toward things Iranian, although Americans had never identified particularly closely and positively with Iran before the fall of the Shah. Because of their emphasis on religious belief as opposed to atheistic Communism, some varieties of Islamic fundamentalism will appeal to Americans; other varieties, by their rejection of Western materialism, will turn off Americans, and this may yet color U.S.

relations with Pakistan or with many countries in the Arab world. As noted, American identification with Israel promises at all times to be stronger than identification with the Arab states hostile to Israel.

Revolutionary possibilities in Central America and Mexico are likely to have an impact also on how Americans feel about these parts of the world. At the best, the countries of Latin America have been seen as duplicates of the U.S. model, fellow republics of the Western Hemisphere, accomplishing their independence of Europe. At the worst, however, they have been seen as repeatedly failing to live up to the U.S. model of constitutional democracy, retreating instead to the autocratic power patterns of the past, or impatiently racing into Marxist versions of the future.

This is all basically to renew the conclusion stated above, that Western (and Eastern) Europe will continue to occupy a special place in the attentions and affections of Americans, when compared to Pakistan or Korea or Honduras or Burma, because the European case will seem the most similar and recognizable of world situations, generating the causes with which Americans can most easily identify. Lech Walesa in Poland will seem like a definite force for good, by American standards, even when such a force cannot be found in the Middle East or in Central America.

CHANGES IN PRECEDENT

As noted, it is difficult to erase the obligations of precedent once they have been established. For a previous U.S. administration to have successfully defended an American position encumbers subsequent administrations with the burden of defending the same position. Old precedents can sometimes be brought to life by the flow of events. For example, if Communist insurgencies now threaten the administration of Corazon Aquino in the Philippines, it will not be as easy for the United States to shrug off the possibility of one more Communist regime's being established in the Southeast Asian region because this is a country that once had the benefits of American rule, where the institutions of political democracy should have had a better chance of being implanted.

It is possible, moreover, that some such additional encumbrances will now emerge in the world because the forces of democracy have been winning some victories against Communist dictatorships around the world. If Soviet forces were to return to Afghanistan, for example, it would be very difficult for the United States to shrug this off, after the Mujahadeen had seemed to win their victory. If the Soviets were to turn away from *glasnost* and *perestroika* or if the Chinese government is now too severe in suppressing student dissent, similar kinds of implicit obligations for protest and measured retaliation may be imposed on the U.S. government. Finally, to return once again to Eastern Europe, if the freedom accorded to Hungary, Poland, and Czechoslovakia (or to Latvia, Lithuania, and Estonia?) were to be compromised, with or without violent Soviet military intervention, such acts would again raise issues of precedent, whereby the United States cannot let such things pass lest its image of power be seen all around the world to have been eroded.

ECONOMIC CHANGES

Thus far we have emphasized the altruistic and the power-oriented interpretations of American foreign-policy commitments. One cannot be generous in this world unless one has some power and influence. But it would be a mistake to see Americans as seeking only after power and influence, for many American efforts have been channeled to better the lives of other peoples.

A Marxist interpretation would instead stress economic competition and economic disputes, as states worry about unemployment and about their economic futures. By this interpretation the United States, rather than being unusually generous, might (as the most capitalist country in the world) have been the biggest troublemaker, continually fighting for access to markets and so on.

It is obvious that this author does not find such arguments very persuasive. If the United States had wanted to be able to dump its exports abroad, it should have allowed Japan to be governed by Stalin after World War II, for this would have very nicely eliminated some major economic competition. A great number of American foreign-policy commitments, ranging from the Marshall Plan to the substantial U.S. contributions to North Atlantic Treaty Organization (NATO) military defense since 1949, would be very difficult to explain by the selfish needs of American capitalism.

It would also be futile to deny that economic variables and economic complications play a major and increasing role in the foreign policies of all the powers. To repeat what has become a cliché, we live in an increasingly interdependent world, where indeed independence requires interdependence. It is becoming impossible for a country to pursue autonomy or autarchy in its economic arrangements because everyone's public will now demand substantial economic growth, and trade and economic specialization seem the only way to achieve this growth.

As one of the largest economic units in the world, the United States is less dependent on such international economic exchange, and somewhat more self-sufficient in sorting out its own economic future, certainly compared to most of the smaller countries of the world, which constantly must be worrying about the next round of trade exchanges. Even for the United States, however, trade, and economic variables in general, is now more important than ever. Increased trade with Japan and with the other rapidly growing economies of the Far East, with the so-called Asian tigers, will affect American considerations of defense policy and foreign policy.

Yet the impact on military factors may not be so profound for at least two reasons. The United States will indeed have trade disputes with Japan and many other countries, but (despite what the Marxist analysis would have projected) these are very unlikely to cause military disputes. The reason that the United States fought World War II with Japan was not that the Japanese (as today) were beating the United States out of customers by more energetic trading practices, but because they were forcing themselves militarily into China and the rest of Asia. While people sometimes today refer whimsically to Japanese success as at last having established a Greater East Asia Co-Prosperity Sphere, this title was a euphemism in the 1940s for a very militaristic imperialism that did not share prosperity with anyone.

A U.S. State Department official was once questioned by a group of Chinese visiting scholars about an impending dispute over the import of textiles from the People's Republic of China, amid fears that this issue would drastically sour Beijing–Washington relations. The answer of the official was very much on the mark when he said that the United States has textile disputes only with its friends. Trade disputes may indeed now be the norm within the portion of the world that is actively engaged in trade (i.e., the portion that has freed up its economic energies enough to be able to compete in market processes), but the writings of Marx and Lenin to the contrary, this will not produce arms races or military confrontations or wars.

There is a second way in which the economic boom that has gripped East Asia has led to a lessened, rather than enhanced, U.S. military commitment to the region: trade and other economic interactions have indeed made the prospect of war generally less plausible and menacing throughout this area. Two decades ago Americans would have been engrossed in scenarios about wars between China and its neighbors or about a string of dominoes falling outward from Vietnam. However, Beijing has undergone a tremendous transformation since the death of Mao, with an economic liberalization far outrunning anything yet proposed in the Soviet Union, amid new de facto trade relationships between Beijing and Taiwan and between Beijing and South Korea.

The Communist regime in Hanoi remained much more faithful to orthodox Marxist theories on how to run an economy but has fallen far behind all the other developing Asian states, becoming, in effect, an economic basket case, and illustrating very well the price of refusing to join the free-world economy.

A liberal economist would have predicted all along that increases of trade can be a substitute for military commitment, rather than generating a need for more such commitments. Only a doctrinaire liberal believer in free trade would argue that this linkage has been confirmed by all the events before and between the world wars, but the East Asian region now offers the pleasant prospect of rapidly growing economies without a growth in military tensions, indeed, with a drop in military tensions.

Predicting the political future of China and Taiwan, of the two Koreas, or of Japan and the members of the Association of South East Asian Nations may always be a difficult exercise, as economic improvements and liberalizations cannot always be matched up so easily with political improvements and liberalizations. Yet the complications here may be more political than economic, and the economic factor may be helpful for easing military tensions.

The American reactions to the 1989 student demonstrations and violent government reaction in Tiananmen Square illustrate a number of the points we have made. It was surely too pessimistic to conclude that the anger shown by the U.S. Congress and public at these events, and the cutting back of links between the American and Chinese defense establishments, would lead to new military tensions between the United States and China or between China and its Asian neighbors. It was also contrary to the real nature of American national character and national interests to expect that the United States would follow the advice of realpolitik analysts (for example, Henry Kissinger) to assign priority totally to considerations of national power.

If Americans cared only about their own national power and about keeping the Soviets in check, they should even have wanted to keep the extreme Maoists in power in China because the Maoists were the most likely to stay antagonistic to the Russians. Earlier, if the United States had wanted to keep Nazi Germany in check, it should have given Japan a free hand in China, rather than risk that Japan would be induced to become an ally for Hitler, confronting the United States with the dreaded prospect of a two-front war. But Americans have always cared about freedom for the Chinese. It was as inevitable that they would welcome the demise of the Gang of Four as that they would oppose Japanese aggressions into China, and it was just as inevitable that Americans would be very upset by the killings of June 1989 and the subsequent arrests and harassments of dissidents and intellectuals.

THE PROLIFERATION OF WEAPONS

We now turn to changes in the arrays of weapons that are deployed around the world, weapons that can change what happens in the fields of military policy, even apart from any continuity or changes in how Americans feel about a region or about the precedents or economic interactions that apply to the region. Some traditionalists might argue that weapons rather than international disputes cause wars. Yet the spread of some kinds of weapons may make wars much more horrible when they occur or may cause the increased likelihood of wars; and they also affect how willing Americans may be to maintain commitments to a region.

We will consider the possibilities of further proliferation of nuclear weapons because this is still normally seen as the proliferation problem. We shall also address the spread of chemical and biological weapons and the spread of new delivery systems for any and all kinds of warheads, such as new aircraft and missiles and new antiaircraft and antitank missiles. Finally, the trends in naval warfare will be considered.

The American public will be reacting to the realities of weapons that have slipped into the hands of nations around the world, and also to the image of such proliferation.

NUCLEAR PROLIFERATION

Nuclear proliferation remains far and away the most dangerous kind of weapons spread because the damage that would be inflicted in a war between Iran and Iraq, or between India and Pakistan, or between any two other countries would be raised horrendously if such warheads were to be used in place of conventional weapons. This is the impact that is probably the easiest to agree upon.

Somewhat more difficult to sort out is whether such nuclear proliferation increases or decreases the chances of a war breaking out and increases or decreases the likelihood of the United States' deploying troops and weapons of its own into the region and becoming involved in any wars that do break out.

Partisans of nuclear proliferation sometimes argue that this will contribute to the deterrence of wars in a region, as each side is dissuaded from launching wars by the destruction we have just noted, just as the presence of nuclear weapons may have prevented wars that otherwise would have erupted in Central Europe since 1945. By contrast, others would note how such weapons might be used preemptively and preventively between two states, as each side's nuclear delivery systems looked vulnerable to the other's first-strike attack, or as the first state to acquire such weapons in the region was tempted to exploit its window of opportunity before the second state acquired any matching capability.

Similarly some observers would expect the United States to pull its forces and naval vessels out of a region once nuclear weapons had spread, lest such forces become too attractive a target for these new nuclear warheads, or because the United States had become fed up with an ally that had acquired such weapons despite strong U.S. pressures against such proliferation. Yet the counterargument might see the United States willing to ally itself even more closely with a state that had acquired nuclear weapons (or especially with states that had simply moved onto the threshold), as the only way to ensure that this nth nuclear-weapon state was not going to do something terribly foolish or irresponsible with its new capabilities.

If there were to be a war between two nuclear-weapon states, Americans would very much have to fear the worst that could happen in such a war. The choice of whether to intervene or whether instead to stay out of the region as much as possible would be a very terrible choice, for all the major powers and not just for the United States; and all of this amounts to one more argument against nuclear proliferation (i.e., against allowing such weapons to spread in the first place).

It has to be stressed that most of the states in these regions have their own powerful reasons to avoid such a spread of nuclear weapons to both sides of potential conflicts. It is hardly a totally selfish policy for Washington (or Moscow) to expend energy in trying to discourage this kind of proliferation. Without this local interest in nonproliferation, it would indeed be very difficult, perhaps impossible, for anyone else to head it off. Our predictive problem is that nuclear proliferation may nonetheless occur, even if it proceeds at a much less rapid pace than many of the more pessimistic analysts have forecast in the past.

The United States has lived for decades with rumors of nuclear weapons in Israel, in South Africa, in India, and then in Pakistan. The pattern of U.S. military commitments in the wake of such rumors and possibilities has been neither a clear withdrawal nor a clear enhancement of commitment. These have all been cases where there still was much left to be confirmed, where there still was a fair amount of damage to be done. Thus these have been cases where Washington still desires to exercise leverage; yet the exercising of leverage is always a difficult hand to play, with no clear and easy advice on how to serve the goals upon which we may agree.

CHEMICAL AND BIOLOGICAL WARFARE PROLIFERATION

The spread of capabilities for chemical and biological warfare (CBW) is causing a great deal of concern around the world. Some of the concern is the same as that with nuclear weapons, that is, whether CBW would greatly increase the damage done in any war (most probably it would because biological weapons, especially, could amount to the poor man's H-bomb) and whether CBW would increase or decrease the likelihood of war.

On the latter point there are scenarios where chemical or biological weapons can be used to further an offensive and thus to encourage a war, in particular where the other side is not equipped with such weapons or with any defensive precautions to counter the impact of such weapons. But what is the situation where both sides have such weaponry and where both sides have been equipped with whatever gear and equipment are thought to be needed to protect against some of CBW's impact? Here, it is possible that the offensive will become less feasible and less attractive to either side for a variety of reasons.

The protective gear that has to be worn against such weapons is typically cumbersome and heavy, thereby slowing down any troops so equipped. The application of such weapons has classically been complicated and unpredictable in its impact, with wind and weather playing an important role; whenever one makes an offensive more complicated in the way it has to be planned and executed, one allows for more things to go wrong in the fog of war, and this is also an argument against the attack, against grand offensive sweeps. If chemical weapons had not been outlawed after World War I, so that both sides in Europe in World War II were fairly confident that the other side would not use such weapons, it has been argued that the German blitzkrieg of 1940 would not have been possible because the German Panzer columns would not have been able to move fast enough while having to be prepared against a Western chemical attack.

Planners of NATO defenses often express fears of Soviet chemical-warfare capabilities, in part because the Soviets have more extensively prepared for such a war, and in part because such weapons can somehow be applied quite cleverly to an effective aggression. But the latter premise has to be somewhat in doubt.

When we consider the impact on peace of the spread of CBW weapons to third countries, all of these same complications would have to be acknowledged. Given an unequal proliferation of such weapons, some temporary asymmetries and inequalities will emerge, which might embolden someone to take the offensive and "strike while the iron was hot." One aspect of any arms race that favors the attack is precisely that weapons may spread unevenly, with the side that is ahead feeling that it should begin the war while it is ahead.

But if such weapons spread simultaneously to various antagonists around the world, they may instead complicate and slow down warfare and also make warfare more painful in ways which might again make such wars less, rather than more, likely.

But we are mainly concerned here with the impact of such proliferation on the willingness of the American public to maintain defense commitments and to intervene militarily in any region around the world.

Again, the presence of such weapons may make it difficult to deploy U.S. forces within their reach. While it would be easier to defend against CBW than against nuclear weapons, it might still not be very easy, and the defensive preparations required for troops and equipment might be so cumbersome that a great deal of American tactical maneuverability would be lost in the process.

As with nuclear weapons, CBW might also give a distant country the capability for threatening the retaliatory destruction of cities within the United States, especially port cities, and it might be very difficult to check the hold of every last tramp steamer.

Countries acquiring such weapons, if they are U.S. allies, may alienate the United States in the process, or as in the case of nuclear proliferation, they may pull the United States into a closer alliance relationship as it seeks to moderate the further initiatives that are taken with such capabilities.

Where the countries acquiring such weapons are instead the adversaries of U.S. allies, Americans could again go in either direction. They could rally more to the defense of an ally as the threatened party (perhaps because it would otherwise be so much threatened by its adversary that it had to reach for nuclear or CBW weapons of its own) or instead back out of the conflict and alliance commitments entirely for fear of the damage that could be inflicted on North America by such weapons.

The latter kind of response, a withdrawal in response to benign behavior by a U.S. ally and proliferation by its adversary, would, of course, amount to a very bad precedent in the eyes of the world, suggesting that the United States no longer could be counted on. Where the U.S. ally was itself the guilty party in terms of such proliferation of deadly and destabilizing weapons, an American withdrawal might rather be seen as a positive lesson for all concerned, whereby the United States requires cooperation by its partners and goes ahead with its threats by terminating the partnership when such cooperation does not emerge.

DELIVERY SYSTEM PROLIFERATION

Even harder to head off will be the spread of advanced missile and combat aircraft systems, delivery vehicles that are multipurpose because they can be used with conventional, with nuclear, or with CBW warheads. The economies of scale on the production of such systems and the great temptations of earning hard currency by sales abroad have made arms-control restraints in this field considerably less serious and possible than on nuclear proliferation.

For many delivery systems, moreover, there is now also a proliferation of counters to such systems, as surface-to-air missiles can stop an attack with modern bombers, or antitank missiles can blunt a tank attack, or antiship missiles can counteract someone else's naval acquisitions.

We must note at least one plausible major difference from the other forms of proliferation we have just discussed. As such new conventional weapons systems become more available, the net impact question will pertain mostly to the likelihood of war, rather than to the damage done in such wars—unless, of course, one already presupposes some extensive proliferation of nuclear, chemical, or biological warheads, as just discussed.

Indeed these delivery systems are hardly indispensable to activating the counter-value threats noted above because the destructive potential of nuclear warheads or chemical warheads may be so great that quite ordinary and old-fashioned modes of transportation would be more than sufficient to hold the cities of the world hostage. These kinds of warheads may indeed be indispensable for counter-value attacks, as even the most advanced kind of new delivery system does not pose a great threat to civilian targets if only conventional warheads are used. (The threat of a conventional rocket directed against someone's nuclear power plant might be one exception to this rule, but this amounts to nuclear deterrence without nuclear warheads, because the nation operating the nuclear power plant, in effect, offers a ready-made de facto warhead to its adversary.)

And the impact on whether wars are encouraged or discouraged will not always be so easy to determine. Some kinds of weapons promise to reward whoever has been the first to launch an attack in a crisis, while others instead suggest that one would be better off waiting in a crisis, letting the other side make the mistake of being the first to attack. We might style the former kinds of preemption-supporting weapons as offensive and the latter as defensive; the lesson here would have to be that not all weapons are so undesirable (i.e., that some types might actually have the impact of making wars less likely).

Apart from whether wars are made more likely or not, we still face the question of whether U.S. defense contributions will be required. For example, if effective antitank weapons were to spread to the confrontation between Pakistan and India, or to that between Israel and its Arab neighbors, these might make it less necessary for anyone on the outside to be prepared to race in to help a client. This might also make it less easy for any outside power to intervene against a state that was an adversary. By advocating defensive defense, the United States might be supporting both peace in the region and a reduction of the burdens of American mutual defense.

Where the weapons that spread are instead of a sort that can be used to launch offensives, and which seem the type best used on first strike rather than second strike, the impacts are then the reverse, with the risks of war going up and the implicit needs for U.S. intervention heightened.

Once again, if one simply projects the likely possibilities of weapons proliferation here, the bottom line for the volume of U.S. defense policy commitment and engagement in foreign quarrels comes out mixed, in part suggesting some retrenchment, in part suggesting greater involvement; in the net it might perhaps lead us to predict about as much entanglement as in the past, rather than foreshadowing any definite change.

NAVAL DEPLOYMENTS

The United States at the end of World War II held a dominance over the seas matched previously only by the British dominance after Trafalgar; each had a navy larger than all the other navies of the world combined. In the 1960s and 1970s the Soviets chose to challenge this preponderance by—for the first time for any Russian navy—putting ships into all the seas of the world. At the beginning of the 1980s the Reagan administration, under the influence of its dynamic Secretary of the Navy John Lehman, elected to respond to this Soviet challenge, and also to seek to exploit other opportunities for naval deployment, by expanding to what was labelled a *six-hundred-ship navy*, and by proclaiming various versions of what was a called a *maritime strategy*.

The Soviets, under the economic and other constraints that have pushed Mikhail Gorbachev into *perestroika* and *glasnost*, have reduced the number of their ships deployed on the high seas, and the United States in the Bush administration is also retreating from its former goal of a six-hundred-ship navy. At the same time, one sees India, China, and other states beginning investments in aircraft carriers and other advanced naval vessels. Furthermore, we have now experienced the first naval combat since 1945—in the 1982 South Atlantic War.

All of this allows us to conclude that, while the superpower naval arms race may cool, U.S. naval preponderance will not return to what it was in 1953, and that there will be other significant naval powers in the future besides the two superpowers. The proliferation of new conventional weapons systems noted above will also include antiship systems. The Argentine success with *Exocet* missiles directed against British ships around the Falklands may simply be a sign of what can happen in future battles between a smaller and a larger power. One again has to try to predict whether the net of all these naval developments will be to encourage forward deployments and to encourage making the first strike if a war seems to be threatening, or whether it will discourage such ventures, rewarding the side that sits still and the side that has the closest land bases. If the latter, the United States will again feel less need, and less opportunity, to deploy naval force around the world. If the former, the need for deployments may be enhanced.

About the Authors

JOHN J. WELTMAN has been a staff member at the Center for National Security Studies at the Los Alamos National Laboratory since 1985. In 1991 he will be visiting scholar at the Paul H. Nitze School of Advanced International Studies, The Johns Hopkins University. He has held numerous academic teaching and research posts in the United States and overseas. He is the author of *Systems Theory in International Relations* (Lexington Books, 1973), *Nuclear Weapons Spread and Australian Policy* (Australian National University, 1981), and editor of *Strategic Defenses and International Stability* (Los Alamos National Laboratory, 1988).

ROBERT W. TUCKER is Emeritus professor of American diplomacy, the Paul H. Nitze School of Advanced International Studies, The Johns Hopkins University. He has lectured widely on issues of American foreign policy and has served as consultant to the State and the Defense Departments. His two most recent books are *The Nuclear Debate: Deterrence and the Lapse of Faith* (Holmes and Meier, 1985) and *Empire of Liberty: The Statecraft of Thomas Jefferson* (Oxford University Press, 1990).

WALTER LAQUEUR is chairman of the International Research Council of the Center for Strategic and International Studies in Washington, D.C., as well as director of the Institute of Contemporary History in London. He is the editor of the *Washington Quarterly* and has been the editor of the *Journal of Contemporary History* since 1965. Among his books are *Europe Since Hitler* (Penguin, 1982), *Russia and Germany* (Transaction, 1990), and *The Long Road to Freedom: Russia and Glasnost* (Scribner's, 1989). He is also co-author of *European Security in the 1990s: Deterrence and Defense after the INF Treaty* (Plenum, 1990), one of the titles in the Issues in International Security book series.

STEVEN A. MAARANEN is Deputy Director of the Center for National Security Studies at the Los Alamos National Laboratory. A political scientist, he has been involved in studies of nuclear strategy and arms control at the Laboratory for more than ten years. He has served as Chief of the Defense and Space Division at the U.S. Arms Control and Disarmament Agency (ACDA), and as acting Deputy Assistant Director of ACDA for Strategic Programs. He is coeditor of *The Future of Conflict in the 1980s* (Lexington Books, 1982) and *Strategic Responses to Conflict in the 1980s* (Lexington Books, 1984).

MICHAEL M. MAY is Director Emeritus of the Lawrence Livermore National Laboratory. He served as a representative on the Threshold Test Ban Treaty negotiating team in Moscow and as a member of the U.S. delegation to the Strategic Arms Limitations Talks. In addition to his position as Adjunct Professor at the University of California, San Diego, he is a member of the Committee on International Security and Arms Control of the National Academy of Sciences and has recently been appointed a member of the Secretary of Energy Advisory Board of the U.S. Department of Energy.

MICHAEL NACHT is professor of public policy and Dean of the School of Public Affairs of the University of Maryland. He has been an aerospace engineer with NASA and has also been a management and systems consultant. He has taught at Harvard University and served as associate director of the Center for Science and International Affairs at the Kennedy School of Government. He is the author of *The Age of Vulnerability: Threats to the Nuclear Stalemate* (Brookings, 1985).

LYNN E. DAVIS is currently a Fellow at the Foreign Policy Institute of the Paul H. Nitze School of Advanced International Studies, The Johns Hopkins University. She has been director of studies at the International Institute for Strategic Studies (IISS) and editor of the Institute's journal *Survival*. She has also served as Deputy Assistant Secretary for Policy Planning in the Department of Defense during the Carter administration, and she has taught at Columbia University and the National War College. Among her publications are *The Cold War Begins, Soviet-American Conflict over Eastern Europe* (Princeton University Press, 1974) and *Limited Nuclear Options: Deterrence and the New American Doctrine* (Adelphi Papers, IISS, 1976).

HARRY HARDING is a Senior Fellow in the Foreign Policy Studies Program at the Brookings Institution. Before assuming his present position, he taught at Stanford University. He is the editor of *China's Foreign Relations in the 1980s* (Yale University Press, 1984) and *Sino-American Relations, 1945-1955: A Joint Reassessment of a Critical Decade* (Scholarly Resources, 1989). He is also the author of *Organizing China: The Problem of Bureaucracy, 1949-1976* (Stanford University Press, 1981), *China's Second Revolution: Reform after Mao* (Brookings, 1987), and *China and Northeast Asia: The Political Dimension* (University Press of America, 1988). He is a member of the advisory committee to the Center for National Security Studies at the Los Alamos National Laboratory.

STEVEN R. DAVID is an associate professor of political science at The Johns Hopkins University, where he directs the International Studies Program. He is the author of *Third World Coups d'Etat and International Security* (Johns Hopkins University Press, 1987). In his forthcoming book, *Choosing Sides: Third World Alignment and Realignment* (Johns Hopkins University Press, 1991), he argues that traditional theories of international relations are inadequate for making Third World alignment decisions understandable and proposes an approach that better explains Third World alignment and foreign policy decisions.

LEWIS A. DUNN is assistant vice president and manager of the Negotiations and Planning Division, Science Applications International Corporation. He has served as assistant director of the U.S. Arms Control and Disarmament Agency and as U.S. ambassador to the 1985 Third Review Conference of Parties to the Treaty on the Nonproliferation of Nuclear Weapons. He is the author of *Controlling the Bomb* (Yale University Press, 1982) and coeditor of *Arms Control Verification and the New Role of On-Site Inspection* (Lexington Books, 1989).

MICHAEL KREPON is president of the Henry L. Stimson Center, Washington, D.C. He previously worked at the Carnegie Endowment for International Peace, the U.S. Arms Control and Disarmament Agency, and the U.S. House of Representatives, assisting Congressman Norm Dicks. He is the author of *Strategic Stalemate, Nuclear Weapons and Arms Control in American Politics* (St. Martin's, 1984), *Arms Control in the Reagan Administration* (University Press of America, 1989), and coeditor of *Verification and Compliance: A Problem-Solving Approach* (London, Macmillan, 1988). His most recent book is *Commercial Observation Satellites and International Security* (St. Martin's, 1990).

JOHN ZYSMAN is a professor of political science at the University of California, Berkeley, and codirector of the Berkeley Roundtable on the International Economy. He has served as consultant to a number of governments and firms in Europe and to firms in Japan and the United States. He is the author of *Governments, Markets, and Growth* (Cornell University Press, 1983), coauthor of *Manufacturing Matters: The Myth of the Post-Industrial Economy* (Basic Books, 1989), and coeditor of *Politics and Productivity: How Japan's Development Strategy Works* (Ballinger, 1989).

GEORGE H. QUESTER has been chairman of the Department of Government and Politics at the University of Maryland, where he teaches courses on defense policy and arms control, and American foreign policy. He previously was chairman of the Cornell University Department of Government. Among his books are *American Foreign Policy: The Lost Consensus* (Praeger, 1982) and *Offense and Defense in the International System* (Transaction, 1987). He is a member of the International Institute for Strategic Studies and the Council on Foreign Relations.

Index

Accidental Launch Protection System, 208–209
Acheson, Dean, 159
Adelman, Kenneth, 101, 102
Aegis cruiser, 202
Afghanistan
  Soviet invasion of, 2, 135, 136
  Soviet withdrawal from, 27, 29–30, 31, 46, 51, 154, 165
  U.S. policy toward, 243
Africa. *See also* individual countries
  decolonization, 11–12
  mineral exports, 171–172
  U.S. policy toward, 3
Alfonsin, Raul, 187
Andropov, Yuri, 41, 103
Angola
  Cuban involvement in, 165
  Cuban withdrawal from, 154
  Soviet alignment, 100, 101, 162, 165, 174
  U.S. policy toward, 172
Antarctic Treaty of 1959, 96, 97, 211
Anti-Ballistic Missile Treaty (ABM), 203, 210
  deterrent aspect, 60, 65–66, 206
  offensive aspect, 72
  Reagan administration adherence, 209
  Soviet adherence, 79
  U.S. domestic policy implications, 97–98
Antisatellite weapons (ASAT), 65, 205, 209–210
Aquino, Corazon, 243
Arab-Israeli War of 1973, 100, 193–194

Argentina
  advanced weaponry capability, 165, 186–187
  as conventional arms exporter, 186–187
  Falkland Islands War, 240, 252
  Soviet policy toward, 53
Armenia, 155
Arms control, 57–75, 93–106, 107–129. *See also* Anti-Ballistic Missile Treaty; Conventional Forces in Europe; Intermediate-Range Nuclear Forces; Strategic Arms Limitations Talks; Strategic Arms Reduction Talks
  for Asia–Pacific region, 147–149
  context of, 58–60
  definition, 94
  future nuclear force reductions, 68–73
  goals, 59–60, 95
  inspection/verification provisions
    by satellite monitoring, 211–212, 214–217
    Soviet cooperation, 27–28
  multilateral, 21
  during 1960s, 95–98
  during 1970s, 98–101
  during 1980s, 101–105
  nuclear parity and, 58–59
  Soviet proposals, 63
  space arms control. *See* Antisatellite weapons (ASAT); Strategic Defense Initiative (SDI)
  "stabilizing reductions" goal, 63–64

257

Arms control (*cont.*)
  strategic implications, 22
  theater nuclear arms negotiations, 66–67, 73
Arms race
  East Asian, 144
  naval, 251–252
  Soviet-NATO, 43–44
Arms sales, to Third World, 33–34, 161, 162, 180
Asia. *See also* East Asia; South Asia; Southeast Asia; individual countries
  decolonization, 11–12
  industrial production, 226
Asian Development Bank, 137
Asia–Pacific Economic Cooperation Conference, 150
Association of Southeast Asian Nations (ASEAN), 133, 141, 161, 150, 246
Austria–Hungary, 6
Automation, flexible, 222–223
Averintsev, Serge, 45
Ayatollah Ruhollah Khomeini, 101, 167
Azerbaijan, 155

Ballistic missiles. *See also* Anti-Ballistic Missile Treaty; specific types of ballistic missiles
  Accidental Launch Protection System, 208–209
  chemical warheads, 192. *See also* Chemical weapons
  containment strategies, 195–196, 197–198
  intercontinental (ICBM), 69–71
    first-strike survivability, 81, 82–83
    funding, 208
    India's development of, 184
    START limitations, 60–61, 62, 64, 77, 78, 81, 82–83
    Soviet reductions, 207
    strategic arsenals reduction, 207
    submarine-launched, 60–61, 62, 69–70, 77, 118–119, 135, 138
    targeting, 86–87
  Middle Eastern capability, 182–183
  regional conflict implications, 184, 187, 188, 189, 190, 191, 192, 193, 194
  Third World capability, 12, 166, 180–181
  warhead limits, 61, 77
Beirut, U.S. Marine barracks attack, 167
Benin, as Soviet ally, 162

Biological weapons, 46, 166, 181, 248–250
Blitzkrieg, 249
Bombers
  Backfire, 135
  B-1B, 62
  first-strike survivability, 80, 82, 83, 84, 90
  military's development of, 229
  START limitations, 60–61, 62
Brazil
  advanced weaponry capability, 165, 186–187
  as conventional arms exporter, 186–187
  debt, 170
  Soviet policy toward, 53
Brest-Litovsk treaty, 14
Brezhnev, Leonid, 41, 42, 46, 50, 93, 98, 100, 103
Bulgaria, 45
Burt, Richard, 102
Bush administration
  antisatellite weapons policy, 210
  arms-control policy, 59, 94
  nuclear-weapons policy, 59
  Panama invasion, 167
  Strategic Defense Initiative policy, 208
  strategic defense policy, 207, 208
  Third World policy, 34

Cambodia
  interim government, 133–134, 137
  Soviet relations, 136, 137, 162
  U.S. policy toward, 133
  Vietnamese occupation of, 51
Cam Ranh Bay, Soviet naval base, 135, 138, 148
Carter administration
  arms-control policy, 100–101
  SALT negotiations, 99, 205–206
  Soviet policy, 1–2
Central America. *See also* individual countries
  Soviet military bases, 162–163
  Soviet noninvolvement, 31
  U.S. policy toward, 3
Central Command, 2
Central Europe. *See also* individual countries
  "denuclearized," 37
  "neutralized," 38–39
  political changes, 34
  Soviet military presence, 5–6, 7
  Soviet withdrawal, 4, 5, 36
Chad, 159

Index

Challenger space shuttle disaster, 205
Chemical weapons, 248–250
  containment strategies, 195, 196, 199
  regional conflict implications, 184, 188, 189, 190, 192, 194–195
  Soviet capability, 46
  as terrorist weapons, 195,
  Third World capability, 12, 166, 177n., 249
Chernenko, Konstantin, 103
Chernobyl, 209, 213
Cold War, termination, 26–27
Comecon, 49
Committee for the Present Danger, 101
Communist party
  attitude toward West, 54
  East Asian, 135
  French, 45
  Hungarian, 7
  Portuguese, 45
Conference on Disarmament in Europe, 211
Conference on Security and Cooperation in Europe, 20–21, 116
Congo, as Soviet ally, 162
Conventional arms/forces
  exports, 186–187, 198–199
  as nuclear targets, 86
  sales to Third World, 180
  Soviet reductions, 28, 29
  strategic applications, 69, 122–125
  of Third World countries, 167, 180
Conventional Forces in Europe (CFE), 67, 107, 108–116
  confidence- and security-building measures, 115, 116, 126, 129
  defense goal, 68, 110–111
  forward defense strategy, 111–112
  future directions, 114–116
  Germany's acceptance of, 124–125
  U.S. European military presence, 113–114
Conventional war
  Soviet defense perimeter, 163
  Third World's strategic importance, 163
Council of Europe, 20
Crowe, William, 202
Cruise missile
  Missile Technology Control Regime, 195–196
  START limitations, 77
  submarine-launched (SLCM), 61, 77, 119
  warning system, 78, 91
CSS-2 missile, 182, 191

Cuba
  Angolan involvement, 154, 165
  opposition to Soviet foreign policy, 45
  Soviet military presence in, 101, 162
  Soviet relations, 43, 53, 162
  strategic importance, 163
Cuban missile crisis, 240
Czechoslovakia
  German invasion of, 126
  1948 coup, 7
  non-Communist leadership, 50
  Soviet policy toward, 45

Defense and Space Treaty, 60
Defense industries, as nuclear targets, 86
Delivery systems, proliferation of, 250–251
Democracy, in Third World, 173
Department of Defense, 206, 230
Disarmament, 94. *See also* Arms control
Dobrynin, Anatoly, 42
Dollar, decreased value, 221
Drug traffic, 29, 167

East Asia, 131–152. *See also* individual countries
  American policy issues regarding, 143–151
    alliance management, 145–146
    arms control/regional disputes, 147–149
    diplomatic strategy, 150–151
    military strategy, 144–147
  declining Soviet influence in, 22
  newly industrializing countries, 168–169
  strategic environment, 131–143
    economic factors, 138–141
    multipolarity, 131, 141–143
    regional disputes, 131, 132–134, 147–149
    Soviet involvement, 131, 134–138
  U.S. military presence in, 23
  U.S. security structure in, 9–11
Eastern Europe
  "neutralized," 38–39
  political changes in, 34
  Soviet policy toward, 18–19
  Soviet withdrawal from, 28, 34–39, 50
  U.S. policy toward, 243–244
  Warsaw Pact and, 34–35
East Germany
  Communist leadership, 45
  domestic autonomy, 18
  Soviet relations, 50

Index

Economy, multipolar global, 219, 225–228, 231–235
Egypt
  ballistic missile capability, 182
  nuclear-weapons policy, 183
Eisenhower, Dwight D., 95, 96, 211
Emigration
  Russian-Jewish, 100
  from Third World, 167
Estonia, 155
Ethiopia, Soviet relations, 101, 162, 174, 175
Ethnic factors, in national security policy, 241–243
Eurocommunist movement, 100
Europe. See also Eastern Europe; Western Europe; individual countries
  nuclear-weapons strategy, 8, 116–121
  political unification, 38–39, 40
  Soviet policy toward, 39
  technology development, 227–228
  U.S. military presence, 7, 22, 113–114, 125–126. See also United States–European alliance
  U.S. policy toward, 25–26
  unified security policy, 219–220
European Economic Community (EEC), 168–169, 227
  European Free Trade Association interaction, 219
  Germany's membership, 20
  political cooperation within, 108
European Free Trade Association, 219
Europe 92 movement, 227
Exocet missile, 252
Exports
  from Latin America, 232
  to Third World, 168, 169

Falkland Islands War, 240, 252
Far East. See also individual countries
  Soviet policy toward, 51
First strike
  nuclear forces survivability, 80–84
  in regional conflicts, 189, 190, 193
France
  nuclear-weapons capability, 8, 21, 58, 67, 73, 119, 121, 128
  satellite surveillance program, 211–212
  U.S.–European alliance and, 20
F-16 fighter, 180
FSX fighter, 145, 233, 234

General Motors, 225
Germany. See also East Germany; West Germany
  denuclearization policy, 29, 37–38
  economic influence, 38
  European Economic Community membership, 20
  invasion of Czechoslovakia, 126
  invasion of Poland, 15
  manufacturing policy, 222, 223, 224
  NATO membership, 9, 122
  reunification, 18, 19, 108
    Conventional Forces in Europe agreement and, 124–125
    NATO membership and, 9, 122
    Soviet Central European policy and, 5–6
  Soviet pact with, 15
  U.S. nuclear weapons in, 126
  unification, 5
  Western European relations with, 20–21
Glasnost, 179
  observation satellites and, 213
  Soviet diplomacy and, 46–47
  Soviet foreign policy and, 42, 54
  Third World policy and, 52
Gorbachev, Mikhail
  arms-control policy of, 103–105
  foreign policy of, 43–55
    criticism of, 44–45, 54–55
    diplomatic style, 46–47
    economic factors, 45–46, 47–49
    Sino-Soviet relations, 50–52
    Third World relations, 52–53
    United States relations, 53–54
  nuclear-weapons reduction proposal of, 63
  popularity, 47
  U.S. support for, 30–31
  World Court proposal of, 29
Greater East Asia Co-Prosperity Sphere, 245
Greenhouse effect, 168
Green Party, 49
Grenada, U. S. invasion of, 167
Gromyko, Andrei, 41–42
Gross National Product (GNP)
  Europe, 159
  India, 169
  Third World, 156
  United States, 226
Gulf of Mexico, Soviet military bases in, 155–156
Gulf War. See Iran–Iraq war

Index

Hague Conferences, 94
Haig, Alexander, 102
Harriman, Averill, 96
Honda, 233
Hong Kong
  economic growth, 138
  exports, 168–169
  return to China, 134
Hot Line agreement, 96
Hungary, 7, 50
Hussein, Saddam, 183

ICBM. *See* Ballistic missiles, intercontinental
Immigration, 241–242
Imports
  price elasticity, 221
  from Third World, 168, 169
India
  advanced weaponry capabilities
    ballistic missiles, 191
    chemical weapons, 184
    nuclear weapons, 165, 184, 185, 188, 197
  foreign policy, 141–142
  Gross National Product, 169
  Soviet relations, 135, 136, 138, 141
  territorial conflicts, 142, 149, 193–194
Indochina
  regional conflicts, 131
  U.S. policy toward, 133
  Vietnamese expansionism in, 14
Indonesia
  economic relations, 140
  territorial conflicts, 142
Industry, as nuclear target, 86, 87, 88
Intermediate-Range Nuclear Forces (INF), 58, 64, 67, 73, 74, 102, 211
  deterrent aspects, 60, 68
  on-site verification provisions, 78
  Reagan administration and, 3, 211
  Soviet foreign policy and, 54
  Soviet relations and, 49
International Atomic Energy Agency, 211
International Court of Justice, 29
International Energy Agency, 170–171
International Monetary Fund, 48, 227
International Satellite Monitoring Agency, 211–212
Iran
  advanced weaponry capabilities
    ballistic missiles, 184

Iran (*cont.*)
  advanced weaponry capabilities (*cont.*)
    chemical weapons, 188
    nuclear weapons, 165–166
  revolution in, 2, 242
  Soviet policy toward, 30
Iranian hostage crisis, 101
Iran–Iraq war, 166, 181, 182, 190, 191, 194
Iraq
  advanced weaponry capabilities
    ballistic missiles, 182
    chemical weapons, 166, 177n., 183, 188
    nuclear weapons, 165–166, 183, 191–192
  invasion of Kuwait, 167
  nuclear research reactor, 182
Islam, 242–243
Isolationism, 159, 237
Israel
  ballistic-missile capability, 194, 202
  chemical-weapons capability, 183
  military assistance to South Africa, 186
  Soviet relations, 28, 161
  U.S. relations, 240–241, 243
Issraelyan, Viktor, 46
Italy
  manufacturing policy, 222, 223, 224
  satellite surveillance program, 212

Jackson, Henry, 101
Japan
  defense expenditures, 10–11, 14, 140, 145, 235
  economic importance, 219, 227
  exports, 168–169
  foreign policy, 141, 142, 143
  international political influence, 16
  Liberal Democratic Party (Japan), 227
  rearmament, 10, 11, 145
  Soviet policy toward, 47, 53, 135, 136
  technology, 139, 221, 226, 234
    capitalist economy and, 222
    military, 220, 233
    production innovations, 223, 224
    spin-on, 228
  territorial conflicts, 142, 149
  as U.S. economic competitor, 12, 244, 245
  U.S. military presence in, 10, 144
  U.S. national security importance, 239
Jericho missile, 182
Johnson administration, arms-control policy, 95–97

Kennedy administration, arms-control policy, 95–96
KGB, 46
Khadhafi, Muammar, 54, 159
Khmer Rouge, 134
Khomeini, Ruhollah, 101, 167
Kissinger, Henry, 98, 99, 100, 246
Kohl, Helmut, 48, 124
Korean Airline Flight 007, 149
Korean Peninsula. *See also* North Korea; South Korea
  advanced weaponry proliferation, 185–186
Korean War, 7, 131, 132, 133, 240
Kosygin, Aleksei, 97
Krushchev, Nikita, 211
Kurds, 166
Kuwait, Iraqi invasion of, 167
Kvitzinsky, Yuli, 102

Lance missile, 121
Laos, as Soviet ally, 136, 162
Latin America. *See also* individual countries
  exports, 232
  Soviet relations, 53
  U.S. relations, 243
Latvia, 155
League of Nations, 237
Lehman, John, 251
Lenin, Vladimir, 14–15
Liberal Democratic Party (Japan), 227
Libya
  advanced weaponry capabilities
    ballistic missiles, 182, 183, 191
    chemical weapons, 166
    nuclear weapons, 165
  American investments in, 172
  arms purchases, 180
  Chad and, 159
  Soviet relations, 28, 30, 54, 162
Ligachev, Yegor, 45
Limited Test Ban Treaty, 96
Lithuania, 155
London Naval Treaty, 94

Malaysia
  economic status, 138
  territorial conflicts, 142
Manufacturing. *See* Technology
Marxism–Leninism, in Third World, 173
McNamara, Robert, 97, 99

Media, observation satellite use by, 212–214
Mercedes, 233
Mexico
  debt, 170
  exports, 169
  imports, 169
  internal stability, 167
  Soviet relations, 161
  U.S. relations, 243
Middle East. *See also* Persian Gulf; individual countries
  advanced weaponry proliferation, 182–183
  Soviet policy toward, 28, 30
Midgetman missile system, 62, 63, 89
Military budget
  Japan, 235
  U.S., 176
  U.S.S.R., 28, 30, 45–46
Minerals, strategic, 171–172
MIRV, 99, 101
Missile(s). *See also* Ballistic missiles; specific types of missiles
  short-range, 77, 120
  tactical air-to-surface standoff, 119
Missile launchers
  START limitations, 62
  Third World capability, 181
Missile Technology Control Regime, 195–196, 197–198
M-9 missile, 185
Mongolia
  Soviet relations, 134–135, 136
  Soviet withdrawal from, 28, 137, 147
Morocco, 163
Mozambique, 162
Multiple, independently targetable reentry vehicles (MIRV), 99, 101
Multipolarity, 17, 131, 141–143
  economic, 219, 225–228, 231–235
MX missile, 103. *See also Peacekeeper* missile

National Security Council, document NSC-68, 25
NATO. *See* North Atlantic Treaty Organization
Naval forces. *See also* Submarines
  deployment, 251–252
  nuclear attack survivability, 192–193
  Soviet, 193
  of Warsaw Pact, 115

Index

Nicaragua
  Soviet relations, 53, 101, 162
  strategic importance, 163
  U.S. policy toward, 3
Nike missile, 185
Nitze, Paul, 100–101, 102
Nixon administration
  arms-control policy, 93, 97–100, 206
  strategic defense policy, 207
Nonproliferation of Nuclear Weapons Treaty, 96, 197, 201
Noriega, Manuel, 175
North Atlantic Treaty Organization (NATO)
  arms-control position, 120
  conventional forces, 66, 73
    reductions, 28, 94, 110. *See also* Conventional Forces in Europe
  European unification and, 219–220
  forward defense strategy, 111–112
  Germany's membership, 9, 122
  nuclear deterrence and, 91
  nuclear-weapons capability, 73, 117
  nuclear-weapons policy, 66–67, 127
  political role, 108, 122
  purpose, 121–122, 129
  Soviet arms race with, 43–44
  Soviet policy and, 20, 54
  theater nuclear forces, 66–67
North Korea
  invasion of South Korea, 159
  nuclear-weapons capability, 134, 185–186, 188, 192
  Soviet relations, 134–135, 136, 162
  U.S. policy toward, 133
North Vietnam, Soviet relations, 134–135, 136, 138
North Yemen, ballistic missile capability, 183
Nuclear export controls, 196
Nuclear-free zones, 58, 138, 148–149
Nuclear freeze movement, 102
Nuclear proliferation, 165–166, 247–248
  containment strategies, 195–198, 200–203
  regional conflict implications, 183–191
    first strike, 189, 190, 193
    superpower involvement, 191–193
Nuclear warheads, START limitations, 61, 77
Nuclear weapons. *See also* Ballistic missiles
  conventional arms versus, 163
  delivery systems, 250–251
  deterrent role, 21, 57, 58, 91

Nuclear weapons (*cont.*)
  first strike
    nuclear forces survivability, 80–84
    in regional conflicts, 189, 190, 193
  land-based, 118–119, 121
  Nonproliferation Treaty, 96, 197, 201
  short-range, arms control, 57, 67
  targeting/target coverage, 21, 84–88, 90
  as terrorist weapons, 195
  testing of, 93
    Joint Verification Experiments, 211
  Third World capability, 12, 181
  U.S. policy regarding, 57–74
    arms-control implications, 57–74
    Bush administration policy, 59
    deterrent objectives, 57, 58, 64, 65
    Reagan administration policy, 59, 69
    reductions, 21–22
    stability objective, 59–60
    strategic defenses and space aspects, 60, 64–66, 72
Nunn, Sam, 208

Oil
  embargo, 170–172
  world price of, 48
Oman, strategic importance, 163
Organization for Economic Cooperation and Development (OECD), 171–172
Organization of Petroleum Exporting Countries (OPEC), 156
Outer Space Treaty of 1967, 65, 97

Pacific area, Soviet policy toward, 47
Pacific Economic Cooperation Conference, 137, 150
Pacific Trade and Development Conference, 150–151
Pakistan, 141, 142
  advanced weaponry capabilities
    chemical weapons, 184
    nuclear weapons, 185, 188, 196–197, 199
  arms purchases, 180
  U.S. relations with, 242–243
Palestine Liberation Organization (PLO), 162
Palestinians, in Saudi Arabia, 171
Panama
  Noriega and, 175
  strategic importance, 163
  U.S. invasion, 167

Paris Peace Conference, 133
Paxsat, 212
Peacekeeper missile, 62, 63
People's Republic of China
  economic status, 138, 139, 140
  foreign policy, 141–142
  missile export controls, 198
  nuclear-weapons capability, 21, 58
  popular protest repression, 51, 134, 141, 246
  Soviet relations, 10, 28, 47, 50–52, 134, 135, 136
  Taiwan's relations with, 134
  territorial conflicts, 142, 149
  U.S. relations, 10, 133, 143, 242, 245, 246
Perestroika, 104, 179
  Asian economic relations and, 137
  Soviet foreign policy and, 42, 47, 54
  U.S. economic assistance and, 31
Perle, Richard, 102–103
Persian Gulf
  American military presence, 144–145
  chemical warfare, 192
  as oil source, 170–171
  U.S. policy toward, 25–26, 175
  as war zone, 192, 193
  Soviet policy toward, 161
Philippines
  Communist insurgencies, 243
  strategic importance, 163
  territorial conflicts, 142
  U.S. military bases, 138, 148
Phillips (telecommunications company), 228
Pipes, Richard, 101
Poland
  German invasion, 15
  Solidarity leadership, 50
Ponomarev, Boris, 42

Rapid Deployment Force, 2
Reagan, Nancy, 103
Reagan, Ronald, 53
Reagan administration
  antisatellite-weapons policy, 205, 209–210
  arms-control policy, 94, 102–103
  East Asian policy, 146–147
  Grenada invasion, 167
  naval policy, 251–252
  nuclear-weapons policy, 59, 69
  Soviet relations, 2–3
  Strategic Defense Initiative (SDI), 205–209

Reagan Doctrine, 165
Red Army, 5–6, 18, 19
Revolution
  in Iran, 2, 242
  in Latin America, 243
  in Third World, 174
Rostow, Eugene, 101, 102
Rowny, Edward, 102
Rushdie, Salman, 43
Ryzhkov, Nikolai, 45

SALT. *See* Strategic Arms Limitation Talks
Sandinistas, 53, 101
Sarney, Jose, 187
Satellite(s), 209–217
  for arms-control monitoring, 211–212, 214–217
  commercial, 212–214, 216–217
  military, 209–212
Satellite Pour l'Observation de la Terre, 212, 213, 214
Saudi Arabia
  advanced weaponry capability, 182, 183
  arms purchases, 180
  importance to U.S., 159
  oil exports, 171
Scowcroft Commission, 103
Scud missile, 166, 182–183, 185
Semiconductor industry, 228–229, 230
Semyonov, Vladimir, 98
Shevardnadze, Eduard, 42, 43, 53
Shmelyov, Nikolai, 48
Shultz, George, 102
Siemens (telecommunications company), 228
Singapore
  economic status, 138
  exports, 168–169
Single European Act, 108
Smith, Gerard, 98
Social Democratic Parties (Europe), 49
Solidarity, 50
South Africa
  advanced weaponry capability, 186
  American investments in, 172
  Israeli military assistance to, 186
South Asia, advanced weaponry capability, 184–185
South Asian Association for Regional Cooperation, 150
Southeast Asia, nuclear-free zones, 138, 148–149

Index

Southeast Asia Treaty Organization, 132
South Korea
  advanced weaponry capability, 185–186, 188
  economic relations, 140
  exports, 168–169
  foreign policy, 141
  invasion of, 159
  trade with China, 245
  U.S. defense of, 132, 133, 143, 144, 145, 159
South Pacific Forum, 150
South Pacific Nuclear Free Zone, 58
South Vietnam, collapse of, 133
South Yemen
  ballistic missile capability, 182
  as Soviet ally, 162
Space
  arms control, 60, 64–66, 72
  antisatellite weapons, 65, 205, 209–210
  commercial uses, 212–214, 216–217
  "open skies" policy, 95, 211–212
  satellites, 209–217
  South Korean space program, 185
  Strategic Defense Initiative (SDI), 65, 69, 103, 205–209, 235
    funding, 206–207
    purpose, 2
    Strategic Arms Reduction Talks and, 207–208
Stalin, Joseph, 15, 42, 44, 238, 244
START. *See* Strategic Arms Limitation Talks
Star Wars. *See* Strategic Defense Initiative (SDI)
Stealth bomber, 233
Straits of Hormuz, 193
Strategic Arms Limitations Talks (SALT), 57–58, 60–64, 74, 76, 109
  I, 58–59, 97–98, 99–100, 211
  II, 58–59, 100–101, 205–206
  during Johnson administration, 97
  NATO and, 66, 67, 73
  during Nixon administration, 97–98
  on-site verification provisions, 78–79
  strategic defense implications, 68, 77–78, 79–91
    civilian fatalities, 88–89
    force survivability, 79, 80–84, 89
    target coverage, 80, 84–88, 89
  Strategic Defense Initiative and, 207–208
  terms of, 60–63
  U.S.–Soviet agreements, 77

Strategic Defense Initiative (SDI), 65, 205–209, 235
  Bush administration and, 206, 208
  funding, 206–207
  nuclear superiority and, 69
  purpose, 2
  Reagan administration and, 65, 69, 103, 205–206, 209
  Strategic Arms Reduction Talks and, 207–208
Submarine-launched cruise missile (SLCM), 61, 77, 119
Submarines
  ballistic missile-launching, 61, 62, 118–119, 135, 138
  cruise, 61, 77, 119
  Minuteman, 62
  Posiedon, 62
  Trident, 62
  first-strike survivability, 80, 81, 82, 83, 84
Su-24 fighter-bomber, 180, 191
Syria
  chemical-weapons capability, 166
  as Soviet ally, 28, 30, 54, 162

Taiwan
  advanced weaponry capability, 186
  American withdrawal from, 133
  China's relations with, 134, 245
  economic status, 138, 139, 140
  imports, 168–169
  industrial production, 226
  mutual defense treaty, 132
  U.S. recognition of, 242
Taiwan Straits, 131, 132, 133, 149
Technology, 219–236
  American decline in, 221–225
    multipolar global economy and, 219, 225–228, 231–235
    security implications, 231–235
  in Asian economic region, 226
  in East Asian economic region, 139–140
  in European economic region, 227–228
  flexible automation, 222–223
  flexible specialization, 223
  location, 229
  military, 220, 233–234
    foreign sources, 232–233
    relationship with commercial technology, 228–231
    spin-on, 220, 228–231

Technology (*cont.*)
  of process control, 223–224
  Soviet acquisition of, 14–15
Terrorism, 29, 47, 167, 195
Thailand
  economic status, 138, 140
  foreign policy, 141
Third World. *See also* individual countries
  advanced weaponry proliferation, 12, 179–202
    ballistic missiles, 166, 180–181
    chemical weapons, 181
    containment strategy, 195–203
    nuclear weapons, 165–166, 181
    regional conflict implications, 179–180, 187–195
    regional trends, 182–187
  arms sales to, 33–34, 161, 162
  autonomy, 17
  conventional forces, 167, 180
  debt, 169–170
  democracy, 173
  emigration, 167
  exports, 168, 169
  gross national product, 156
  imports, 168, 169
  leadership
    economic policies of, 170
    security concerns of, 174
    Soviet support for, 162, 164–165
  nonalignment, 7
  regional conflicts, 160
    advanced weaponry and, 179–180, 187–195
    global involvement, 191–195
    patterns of, 187–191
  revolutions in, 174
  Soviet declining influence in, 22
  Soviet military presence in, 162–163
  Soviet policy toward, 47, 52–53, 154–156, 157, 161–165
  superpower nonalignment, 11–12
  U.S. national security issues, 34, 153–177
    drug trafficking, 167
    economic factors, 156, 168–172, 175–176
    environmental hazards, 168
    immigration, 167
    policy priorities, 175
    political–ideological factors, 156–157, 158, 173

Third World (*cont.*)
  U.S. national security issues (*cont.*)
    Soviet Third World policy, 154–156, 157
    strategic–military factors, 154–156, 158, 159–160, 161–168, 175–176
  U.S.–Soviet relations and, 2–3
Thomson (semiconductor manufacturer), 228
Tiananmen Square, 51, 141, 246
Trade deficit, 221, 232
Treaty for Nonproliferation of Nuclear Weapons, 197, 201, 211
Treaty for the Prohibition of Nuclear Weapons in Latin America, 97
Tropical forests, destruction of, 168

Union of Soviet Socialist Republics
  credit policy, 48–49
  defense perimeter, 163
  foreign policy
    toward China, 10, 28, 47, 50–52, 134, 135, 136
    toward Europe, 6–9, 18
    of external retrenchment, 4–6
    under Gorbachev, 4–6, 41–55
    toward Israel, 28
    recent changes, 27–30
    reversibility, 30–33
    toward Third World, 11–12, 154–156, 157, 161–165
  military budget, 28, 30, 45–46
  military policy, 44
    conventional forces reductions, 28, 29
    in Asia, 51, 131, 134–136, 138, 147
    in Eastern Europe, 28
  Missile Technology Control Regime adherence, 197–198
  naval deployment policy, 251–252
  New Economic Policy, 14–15
  oil exports, 48
  territorial disputes, 149
  trade, 47–48
UNITA, 165
United Kingdom
  European military presence of, 125–126
  nuclear-weapons capability, 8, 21, 58, 67, 73, 119, 121, 128
United Nations, 29, 162
United States
  Chinese policy, 10, 133, 143, 242, 245, 246
  debt, 221, 232

Index

United States (*cont.*)
  declining economic position, 12–13
  Eastern European policy, 36
  European alliance
    ethnic factors, 241–242, 243
    Soviet policy changes and, 6–9, 19–22
  future defense commitments, 237–252
    biological-weapon proliferation and, 248–250
    chemical-weapon proliferation and, 248–250
    delivery system proliferation and, 250–251
    economic factors, 244–246
    ethnic factors, 241–243
    naval deployments, 251–252
    nuclear-weapon proliferation and, 247–248
  gross national product, 226
  military budget, 176
  naval deployment policy, 251–252
  nuclear-weapons policy, 57–74
    arms-control implications, 57–74
    of Bush administration, 59
    deterrent focus, 57, 58, 64, 65, 68, 69, 71, 74
    of Reagan administration, 59, 69
    stability objective, 58–59
    strategic defenses and space aspects, 60, 64–66, 72
  Soviet policy, 25–40
    of Carter administration, 1–2
    economic factors, 13–14
    historical background, 25–26
    ideological factors, 15–16
    of Reagan administration, 2–3
    strategic defense implications, 207
    during World War II, 15
  surplus economic resources, 232
  trade deficit, 221, 232
U-2 flight, 211

Very High Speed Integrated Circuits, 230
Vietnam, 157. *See also* North Vietnam; South Vietnam
  American withdrawal from, 98
  economy, 140, 245

Vietnam (*cont.*)
  expansionism, 14
  foreign policy, 141
  as Soviet ally, 43, 45, 162
  territorial conflicts, 142
Vietnam War, 25
  Paris Peace Conference, 133

Walesa, Lech, 243
Warnke, Paul, 101
Warsaw Pact, 108
  conventional forces, 66, 67, 73
    reduction, 94, 112, 114. *See also* Conventional Forces in Europe
    superiority, 117
  defensive restructuring, 28, 111–112
  Eastern European self-determination and, 34–35
  political stability role, 122
  termination, 38
Washington Naval Treaty, 71, 94
Weinberger, Caspar, 102
West Berlin, 240
Western Europe
  defense position of, 235
  gross national product, 159
  political unification, 16, 38
  Soviet policy toward, 47
  U.S. security importance, 239
Western European alliance, 7–9, 16. *See also* United States, European alliance
Western European Union, 8, 20
West Germany
  exports, 168–169
  Soviet relations, 7–8, 9, 49–50
  U.S. military presence in, 113–114
Wilson, Woodrow, 237
World Bank, 48
World Court, 29
Wyse Technologies, 226

Yalta, 36
Yeltsin, Boris, 104
Yemen, Soviet involvement in, 100, 101

Zaire, mineral exports, 172
Zimbabwe, mineral exports, 172